THE RIVER KILLERS

THE RIVER

Stackpole Books

KILLERS

by Martin Heuvelmans

353.008232
H595

Printed in the U.S.A.

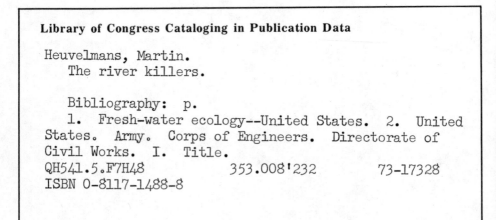

Library of Congress Cataloging in Publication Data

Heuvelmans, Martin.
   The river killers.

   Bibliography: p.
   1. Fresh-water ecology--United States. 2. United
States. Army. Corps of Engineers. Directorate of
Civil Works. I. Title.
QH541.5.F7H48   353.008'232   73-17328
ISBN 0-8117-1488-8

In gratitude to a kind fate
That placed me upon these shores
I humbly dedicate this book to
the preservation of
the beauty, glory and the grandeur of
the United States of America.

Contents

America's environment ruined by the Corps of Engineers. Florida invaded. The Corps' impact on the state's ecology. Hurricane disasters—2750 people die. The Corps' remedies threaten Florida's survival.

ened by plans for the Sacramento River Delta. The Northwest dammed beyond all practical needs. Excess power capacity already exists. The salmon industry ruined. More than 6,000,000 salmon die in one year. Hells Canyon threatened. More chicanery by the Corps. The Corps in trouble.

Chapter 8   ABOLISH THE CORPS   164

The Corps has outlived its usefulness. Authorities condemn the Corps and recommend a complete change of responsibilities. A new voice in the land. American voters become environmentalists. How to break up the partnership between the Corps and Congress. How to save America's remaining water resources. Only one answer.

APPENDIX

Preface

PUBLICATION OF *The River Killers* climaxes an effort which began more than a decade ago with my casual inquiries about the silting pollutions of local waters around Stuart, Florida. Because reasonable answers were not forthcoming, I became more and more engrossed in the methods and operations of the Civil Works Projects of the Army Corps of Engineers, which, to my dismay, were less than honorable.

The destructive nature of Corps' projects has been treated quite effectively by others and is touched on in *The River Killers* only as necessary to support the truth of the Corps' blatant contempt for all but its self preservation, and the misleading, deceitful, and often illegal methods it uses to foist its politically-and-power group-motivated projects on an unsuspecting citizenry.

Being just an ordinary citizen, I do not have personal influence at any level of government, but my research and involvement has proven that some in high places share my views. U.S. Supreme Court Justice Douglas has referred to the Corps as "Public Enemy #1;" former Secretary of Interior, Stewart Udall has written in the public press that the Corps suffers a "jurassic mentality."

There is no question in my mind that the killing of our rivers, and the destruction of large portions of our environment which rely on water in one way or another, is imminent, and can only be saved by such a public outcry as has not been heard in the hallowed halls of Congress in many years.

Citizens and ecological groups are teaming up to battle the Corps on local projects everywhere, but what is needed is a tidal wave of protest that cannot be misunderstood by those who owe their elective office to the majority of the people and not to the special-privilege and power groups who stand to profit in some way from this gross misuse of public money and domain. I sincerely hope *The River Killers* will spark that protest.

Martin Heuvelmans, Citizen

Acknowledgments

ACKNOWLEDGMENT IS GRATEFULLY made to the following. Without their kind assistance, much information and documentation so vital to the researching of this work would have been much more difficult, if not impossible to obtain.

Such assistance as was extended to me by public officials, was given as a matter of the normal responsibility of their office. No position, either con or pro, is attributed or implied.

✿ ✿ ✿ ✿ ✿ ✿ ✿ ✿ ✿

Executive Offices, The President of the United States
Senator Edward J. Gurney
Senator Lawton Chiles
Hon. Paul Rogers, MC
Hon. John D. Dingel, MC
Hon. L. A. "Skip" Bafalis, MC
Hon. Nathaniel P. Reed, Assistant Secretary of the Interior
for Fish, Wildlife, and Parks
Hon. Stewart Udall, Former Secretary of the Interior
Governor Reubin Askew, Florida
Attorney General Robert L. Shevin, Florida
Wellborn Jack, Jr., Esq., Shreveport, La.
Bud Bristow, Chief Project Evaluator, Arizona Fish and Game
Department

Arthur Solomon, Jr., President Northwest Steelheaders Council, Spokane, Washington

Bruce Hannon, Engineering Department, University of Illinois

The Committee on Allerton Park, Champaign, Illinois

John A. Briggs, Director, Department of Ecology, State of Washington

Joseph C. Greenley, Director, Fish and Game Department, Idaho

Jonas Morris, Morris Associates, Consultants for Governmental Affairs, Washington, D. C.

National Parks & Recreation Association

Ted Radke, Ecology Action of Contra Costa, California

Jerry Jane, President, Idaho Environmental Council

Pennfield Jensen, Editor, Clear Creek Magazine

Darrell E. Louder, Director, Fish & Wildlife, Dover, Delaware

Jim Pozewitz, Chief, Environmental Resources, Montana Fish & Game Department

Christopher Percy, Executive Director, Connecticut River Watershed Council

Ozark Society, Fayetteville, Arkansas

National Marine Fisheries Service, U. S. Department of Commerce

Fish and Wildlife Service, Department of the Interior

Arthur R. Marshall, Director, Applied Ecology, University of Miami

Ernest Lyons, Editor, Stuart News, Stuart, Florida

Timer Powers, Martin County Commissioner, Florida

South Central Florida Flood Control District, West Palm Beach, Florida

Florida Game & Fish Commission, Dr. O. E. Frye, Research Director

Florida Board of Conservation, Dr. R. M. Ingle, Research Director

And, finally, to Wesley Marousch of Stuart, Florida for his help in preparing my original manuscript, and to the Publisher and editorial staff of Stackpole Books, especially Tom McElroy, for their confidence in my story and their healing of the wounds of my assaults upon the King's English, my deep and undying gratitude.

Introduction

AMERICA'S HISTORY HAS been molded largely by man's confrontation with his environment. Early settlers found a bounteous land rich in natural resources—a land of vast forests, open prairies, endless waterways, great mineral deposits, and abundant wildlife. The needs of survival necessitated certain environmental changes, but it was personal greed and exploitation that quickly ravished this new land. Today, much of America's virgin timberlands is gone; much of its rich topsoil lies beneath impounded waters and river deltas; its wildlife is depleted, its waters polluted, and its scenic grandeur defaced by the acts of man. The blame for the pathetic blanket of despoliation that spreads across America must, in part, be shared by all its people, but surely, no one group or agency has done more to bring about this national tragedy than the Civil Works Branch of the United States Army Corps of Engineers.

The Corps of Engineers originated in the Continental Army. As President, George Washington established an engineering school at West Point in 1795. In 1802, the present Corps of Engineers and the United States Military Academy were established by President Jefferson. Thus, for more than two decades, West Point was the country's only training center for engineers. Government officials had no recourse but to turn to the Army for the nation's engineering needs.

Congressional legislation in 1824 established a Board of Internal Improvements. This board, consisting of two Army Engineers and one civilian engineer, was assigned the duty of developing a national land and waterway transportation system. This was the beginning of the Civil Works Branch of the United States Army Corps of Engineers. It is to this branch and this branch exclusively—the Civil Works Branch of the Army Corps of Engineers—that this book hereinafter refers to as "the Corps."

As the country grew, the Corps grew with it. The Corps' first assignment was primarily one of improving navigation facilities, but other responsibilities were soon added—flood control, the regulation of hydraulic mining, bridge permits, water pollution, waterway development, irrigation, hydroelectric power, recreation, and other projects allied with the nation's water resources. In order to maintain control over such numerous and diverse interests, the Corps established a nation-wide system of regional and district offices. By 1942, the Corps was spending the taxpayers money at the rate of $20 million per day. Today, with its 200 military engineers and 32,000 civilian employees, it spends 75 percent of yearly water development appropriations—over $1.5 billion.

These responsibilities, and an almost limitless budget, have made the Corps an elite and politically powerful segment of American society. Its strength has been used to perpetuate its self-aggrandizement and to satiate the political whims of some members of its patronizing partner—the Congress of the United States.

Corps projects and their effects upon the environment are generally known, but the methods of pursual employed by the Corps in getting projects approved and financed are lesser known factors. How does the Corps manage to maintain its autonomous position despite strong public opposition? Why and how does the Corps generally get what it wants regardless of environmental damage and public costs? How does it proceed? These are the basic questions answered by this book. And in doing so, it reveals the Corps' reasons and methods of procedure to be manifold—often devious, and sometimes purposely deceitful.

One finds the Corps indicted herein as the killer of America's rivers—not by the author alone, but by the writings and opinions of officials and intellectuals of national prominence. Exhibits of substantiating evidence lie scattered in the waterways all across the nation. The citizens of America sit as judge and jury; their verdict must be one of guilt, and the punishment no less than the abolishment of the Corps. To find otherwise is to sanction the continuing destruction of a finite land.

PART I
SPOTLIGHT ON FLORIDA

CHAPTER 1

Before the Corps Came

THE AMERICAN LANDSCAPE lies tattered and trampled beneath the marching feet of the implacable Corps of Engineers. Trails of irreparable disaster stand as monuments to its passing by.

Devoted to the military ethic that all movement is best in a straight line, rivers have been channelized and dammed, and estuaries smothered in silt. Wetlands have been drained and dry lands flooded. Wildlife habitats have been destroyed without thought of the species involved. Vast areas of scenic beauty, and areas of scientific and educational importance, lie buried beneath the silted waters of the Corps' impoundments. In its determined march across the land, the Corps has left the American people a heritage of ecological and economic tragedies that will be their shameful burden to carry for a long time to come.

And the scheming of the Corps to perpetuate itself has not abated; from coast to coast every segment of the country is within the scope of its ever-searching surveyors' transits. The crunch of bulldozers and the screech of draglines continues to sound a requiem for the American landscape.

Nowhere, as in Florida, have the operations of the Corps had a more devastating impact on so fragile an environment, and nowhere have a people been left with so perilous a future because of these operations. The Corps, itself, now predicts that southern Florida will face a severe water shortage within the next few years. Will it become the first desert in the world with an annual rainfall of sixty inches?

Knowing something of the state's physical characteristics and its ecological pattern is necessary to understanding why the Corps has had such a devastating effect upon Florida's terrain. The semi-tropical area of southern Florida is an infant among the geological formations of the United States. Perhaps no more than 5000 years ago this porous limestone peninsula emerged from a receding sea. Sand, shells, skeletal marine life, and eventually vegetation continued to build the peninsula into a unique topographical region most of which stands alone by ecological definition: here we find the only Everglades in the world.

Florida is surrounded by water, largely immersed in water, and dotted with numerous lakes. Yet, paradoxically, the growing need for a continuous supply of fresh water threatens its very survival. Trace Florida's latitudinal line around the globe, and you see that other lands on this line are mostly desert. The prevailing moist winds of the Atlantic, in conjunction with warm land-thermals, produce an abundance of rain, and it is solely by rain that southern Florida survives. Here, rain is everything.

The water lifeline that sustains the millions of people crowding Florida's southern shores begins with the Kissimmee River watershed in the central part of the state. This river flows southward into Lake Okeechobee. South of the lake, the great "river of grass" known as the Everglades extends some 100 miles to the southern tip of Florida. It reaches fifty miles in width, has an average depth of less than one foot, and flows at a rate of less than one mile per day. Yet, despite these unimpressive statistics, the Everglades form the master link in an ecological chain of events that sustain this lifeline of survival. Basically, the Everglades can be looked upon as a huge, shallow, limestone basin that impounds fresh water so as to exert an equalizing pressure against the ever-threatening intrusion of salt water. If this pressure basin should be lost because of the lack of fresh water, the sea would soon reclaim the land by contamination.

But it is not water, alone, that sustains this lifeline. Myriad species of wildlife, from microscopic forms to alligators and panthers, are found here. The vegetation is lush and varied; it slows and holds the water, prevents excessive evaporation, and provides food and cover for an abundance of life. Water, wildlife, plants—one cannot be destroyed without losing all. Yet, it is this very lifeline—this lifeline, so delicately balanced and so vital to southern Florida—that the Corps has severed, and severed again and again.

Before the Corps came to Florida, the Kissimmee River flowed lazily in serpentine fashion through ninety miles of lush countryside before emptying into the lake. Upon entering the lake, this water, along with that from lesser streams, spread out over a surface area of some 800 square miles. The southern shores of the lake were sufficiently low to disperse excess water in times of storms and floods. In addition, the Caloosahatchee River flowed westward from the lake to Ft. Myers and the Gulf; the St. Lucie Canal flowed eastward to Stuart and the Atlantic.

In the mid-1920's Florida was experiencing a tremendous land boom. People soon recognized the potential value of the rich mucklands surrounding Lake Okeechobee, and settlement on this natural flood plain began in earnest. As the agricultural areas expanded and were developed into prime farming land, levees were built to shut off the periodic overflows from the low-lying southern shores of the lake. Four agricultural canals had been built, and it was thought that these canals, plus the St. Lucie Canal and the Caloosahatchee River, would be sufficient to handle the lake's overflow in times of emergency. This theory proved to be wrong, and despite the spending of millions of tax dollars, has not been improved.

It all began in 1926 when a massive hurricane with extremely heavy rains cut across the state of Florida causing the waters of Lake Okeechobee to rise to great heights. Pounded by devastating winds and high waters, the low earthworks were easily topped by the strong waves. Quick erosion of the levees resulted, and sufficient waters rushed through the breaches to drown 250 people.

The levees were repaired and strengthened during the following year in an effort to prevent a recurrence of this tragedy. It was believed at the time that the new levees were positive protection against future disasters, and people returned to the flood plain in even greater numbers. But it was not long before they realized that these attempts to control the forces of nature were tragically inadequate.

In 1928, the new levees were overwhelmed by a tremendous wave of water that virtually obliterated them. Yet the winds which caused this catastrophic damage were barely of sufficient force to be termed a hurricane. A hurricane with wind of only eighty-four miles per hour crossed Lake Okeechobee from a southerly direction. The waters of the lake were driven toward the north shore by this wind, and then a lethal coordination of forces occurred.

The eye of the hurricane passed directly over the lake. Since the center of a hurricane is a dead calm, the waters that had been piled up against the north shore began to flow back toward the thirty-mile-distant south shore as a small tidal wave. This in itself probably would not have caused much harm. However, after the calm of the eye had passed, high winds started blowing from the north. These high winds picked up the wave travelling southward and augmented its force until it developed into an unimaginable tidal wave. When it struck the south shore it destroyed everything in its path. Water heights of thirty feet above sea level, or about seventeen feet above the normal level of the lake, were recorded. As a result, 2750 people died. Press photographs of the day reveal funeral pyres 10 feet high consuming the bodies of the dead.

Ironically, it was these two successive tragedies, and the suggestion for their remedy, by President Hoover, that opened Florida's "flood gates" to the on-rushing Corps of Engineers. And the Corps really came flooding in! It left in its wake wounds and scars more tragic than nature herself could ever impose upon a landscape.

The Presidential suggestion was that an immense levee be built around Lake Okeechobee, and this became the first step in the greatest earth-moving project the world has seen since the digging of the Panama Canal. Seemingly, the Corps interpreted this project as an invitation for permanent tenure within the state, it has been there ever since, and it shows no signs of leaving despite the shameful fiascos that resulted from its presence. If and when the Corps' projects planned for Florida are ever completed, the ditches it has gouged would be of sufficient length to encircle most of the New England states.

Most of the projects thrust upon the state have been in some way associated with Lake Okeechobee, ostensibly for flood control. True, the threat of flooding to southern Florida has lessened somewhat, but this is actually the by-product of massive land reclamation. With the American taxpayer footing the bill, it has been the land speculators, cattle ranchers, and agricultural interests that have really profited from the Corps' intrusion. And in this grabbing of land, one important factor has been overlooked: the continuous flow of water southward from the lake with periodic flooding is really a natural phenomenon essential to the formation and survival of southern Florida!

With the building of higher levees, agricultural speculators moved in south of the lake and started carving up the northern Everglades. An area nearly the size of Rhode Island was leveled by their plows and harrows. This action, along with other politically approved, self-seeking projects, brought a period of threatening disasters to the Everglades.

But to put things in order, let us consider the ironic chain of events that began to unfold. Recognizing the threat to the Everglades, and the fact that they are uniquely Americana, the government established Everglades National Park. In dedicating this park in 1947, President Truman stated, "Here is land tranquil in its quiet beauty, serving not as the source of water but as the last receiver of it."

Then, in 1948, Congress gave the Corps every requested appropriation, and its agile water boys soon began re-shaping what nature had carefully provided.

By 1949, the Corps was ready to launch its infamous "flood control project" for southern Florida that would eventually cost the taxpayers an estimated half billion dollars.

Then, in 1966, disaster struck again. The mishandling of water, combined with a season of drought, nearly ruined the relatively new national park, and threatened all of southern Florida. Thousands of acres went up in smoke; the stench from dead and dying wildlife filled the air for miles—all for the lack of water. And yet, in the very season when the drought was at its worst, the Corps' ingenious system of ditches carried billions of gallons of fresh water into the ocean.

The Corps' manipulations of Lake Okeechobee's waters were not confined only to the southern areas. Its ditches and canals surround the lake like a giant checkerboard. To the north, the gentle-flowing Kissimmee River became a straight sluiceway half its original length. Here, too, land developers and cattlemen were quick to grab up the newly drained land at virtually giveaway prices. The waste from new cities and from thousands of cattle poured down the sluiceway into the lake at an astounding new rate of speed.

To the east, the St. Lucie Canal has been so mismanaged that the last great estuary on Florida's eastern seaboard lies buried beneath mountains of silt from the Corps' own diggings. The loss of this estuary can be measured by the vanishing commercial fishing industry, by devalued real estate, and perhaps most vividly of all, by the destruction of the breeding or nursing grounds for mollusks, crustaceans, and approximately 80% of the fish species native to Florida's coastal waters.

The Caloosahatchee River, flowing westward from Lake

Okeechobee, has suffered virtually the same fate. Millions of dollars were spent in straightening and widening the river so that it might help carry emergency overflows from the lake. All this digging was an exercise in futility; it accomplished little more than to make it possible for the river to handle the excess runoff from its own watershed.

Farther north of Lake Okeechobee, and completely remote from it, the most inane project ever perpetrated by the Corps lies as a permanent scar across Florida's landscape. Like an open wound that will not heal, the Cross-Florida Barge Canal is an agonizing reminder of the Corps' persistent fallacies. But, painful as it may be, the scar is emblematic of the first successful battle to stop the Corps.

Before the Corps came to Florida, man and nature existed in a relative state of balance. Man built upon land that was suitable for building; he farmed land suitable for farming; and he grazed his cattle where they would not drown. Nature's authority over the land was sometimes harsh, but both man and Florida survived. Now, Florida exists with an apprehensive eye on the future, and justly so, for the parental sea threatens to reclaim her mistreated child.

The Corps' excuses for being in Florida ring with about as much sincerity as a schoolboy's stammering reasons for being tardy. "Flood control" and "authorization by Congress" are the stock explanations—sympathetic appeals to congressmen who are anxious to make a splash with the folks back home. Actually, the initial spark that ignites the Corps' monstrous machines is often a lot less consequential than the reasons given. As with the Kissimmee River, it can be as simple as "someone's cow got wet feet."

Ironically, the Corps perpetuates its own "needs" for its tenure in Florida and elsewhere throughout the country. When an area is drained or a dam is built, "new" land is created, and it is soon crowded with people. These people demand greater protection from the very things the Corps sought to alleviate. More pretentious projects are started which, in turn, attract more people. The cycle continues and grows more calamitous with each move.

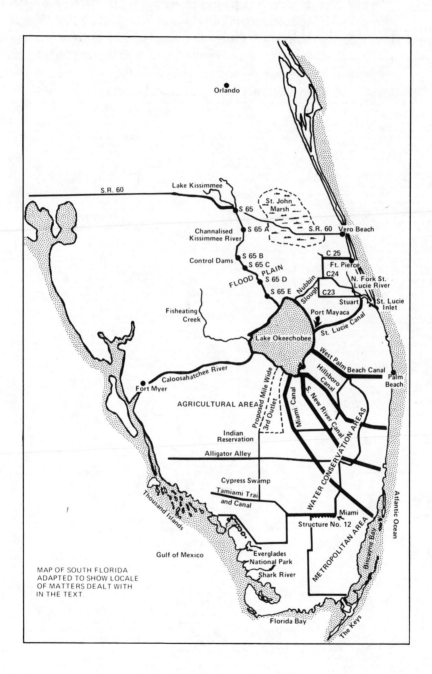

MAP OF SOUTH FLORIDA
ADAPTED TO SHOW LOCALE
OF MATTERS DEALT WITH
IN THE TEXT.

The fact that the Corps has overstayed its welcome every-where, that it has employed devious means to do so, and that its projects are largely failures are documented in the chapters which follow. The visual evidence that it has brought havoc to our land stretches from coast to coast for all to see. America is the victim; America now points an accusing finger.

CHAPTER 2

Blueprint for Disaster

THE RETAINING OF Lake Okeechobee's waters and the building of allied drainage canals enhanced the development schemes that were running rampant. A "new Florida" was being born, and everyone wanted his share of the real estate that was being carved to pieces. Landowners started drainage projects of their own, and canals started radiating across the landscape like strands of a giant spider web—often nullifying the effectiveness of each other and the work of the Corps. One man's dry land became another man's drainage basin. The situation became so chaotic that Congress called upon the Corps to bring order to this confusing situation.

The Corps responded with the *Comprehensive Plan of 1947*. In this plan, the Corps did an excellent job of appraising and evaluating the broad concepts of the problem, and established sound principles for its solution. This massive and rather formidable document, along with its six appendices and accompanying engineering data, remains as the basic guideline today. But as with most far-reaching plans, it was subject to divergent political interests and legal nit-picking. Legitimate flood-control needs had to compete with the politically powerful voices of land speculators. Farmers and cattlemen wanted drainage of wet acreage and flood protection for it; cities insisted on protection against floods; and the conservationists, who wanted all work done in a manner to conserve water for dry times and to protect wildlife habitats, were politely brushed aside as a bunch of radicals who wanted to halt progress. In self-defense, the Corps insisted on a single agency to represent local interests.

❧   ❧   ❧

In response to the Corps' dilemma, the state legislature created the Central and Southern Florida Flood Control District (FCD). The procedural directives for this agency stipulated that local interests must be cleared through it before any action is taken. The FCD would consider the pros and cons of each request and ask the Corps to act on those it considered most urgent.

Seemingly, this new agency, with its simplified methods of procedure, should have ended the haphazard manner in which conflicting interests were carving up the Florida landscape. But in creating the FCD a new body politic was thrust into the middle of the lucrative business of land disposal. Florida's legal bodies now had a magic wand to wave before their constituents.

Much of the land under question belonged to the state. The Florida Cabinet, wearing the hats of the Internal Improvement Fund, supposedly held this land in trust for the people. But in the early days of the FCD, the board was padded with cattle and agricultural interests. These interests came up with the prime acreage from reclaimed land.

Fortunately, within the past ten years or so, this blatant type of pressure politics has disappeared from the FCD's operations. Persistent work by conservationists, environmentally minded journalists, and concerned legal sources have brought about some drastic changes. The board has been enlarged and includes representatives from all facets of interest, including people who are deeply concerned about Florida's environment.

The role of the FCD is not an easy one. There are still diversified interests exerting what pressure they can; there is still pork-barrel money to be spent; and, of course, there is the Corps ready to start digging in the first puddle it finds. The FCD no longer adheres to the Corps' view that the way to handle flood waters is to impound them behind a dam or rush them out to sea in straight-line ditches. Despite the rather implicit connotation of its name, the Flood Control District is now concerned with the wider aspects of complete water management. In fact, it is so charged legally, and this is its only reason for existence.

❧   ❧   ❧

Ill-conceived plans bring disastrous results. Beginning with the St. Lucie Estuary, this obvious fact was to haunt the Corps unmercifully in its sustained march across Florida.

One must have a visual concept of this great estuary (see map, page 31) to understand how such plans and their improper implementation were to completely change the biological complex of so large a segment of Florida's coastal waters. The estuary is formed by the confluence of the North Fork estuary and the South Fork estuary of the St. Lucie River. At this junction the river widens, flows gently to the east and then to the south where it joins the waters of the Indian River and flows into the Atlantic via the St. Lucie Inlet.

Before the arrival of the Corps the estuary was a semitropical paradise that supported multitudinous forms of aquatic and terrestrial life. These life forms existed in an ecological balance which was not understood, nor even considered, in the plans formulated by the Corps.

The upper reaches of the estuary contained eelgrass flats, marshes, mangrove swamps, and adjacent woodlands—each, in its own way contributing to the support of all life from there downstream to the Atlantic. The enriched waters of the marshes in the natural flood plain supported myriads of phytoplankton, unicellular plants that oxygenated the water and provided food for the animal plankton. Top minnows and killifish fed on the plankton; they in turn were good for amphibians, fish, water birds, and other higher forms of life in the estuary's life-support system. Eelgrass flats were alive with shrimp—food for an infinite variety of fishes. The mangrove borders held the soil and provided cover for still other interrelated chains of life.

As the estuary's pyramid of energy continued to build, the larger fishes were supported in great variety and numbers—bluefish, mackerel, trout, channel bass, flounder, snapper, snook, tarpon, and others. And finally, on the pinnacle, was man, for the St. Lucie area was the center of a large sport and commercial fishing enterprise. It was this enterprise, this whole life-support system for the estuary, that the Corps was deigned to destroy in its plans to rush excess waters to the sea.

As a result of the recommendations in the *Comprehensive Plan of 1947*, Congress authorized the funding and the building of the St. Lucie Canal. The canal was to be deepened and widened, and a dam built to control the overflow of fresh water from Lake Okeechobee. The canal empties into the South Fork of the St. Lucie River.

Contingent plans called for two additional canals (C-23 and C-24) that would terminate in the North Fork of the river. These were to be irrigation canals and control outlets that would serve both St. Lucie and Martin Counties. Their source would be the St. Johns Marshes immediately to the north.

The plans, themselves, concealed obvious and inherent dangers to the estuary, but the manner in which they were implemented was the beginning of the Corps' acquisition of that rather ignominious title, "The River Killers."

⚜   ⚜   ⚜

The Kissimmee River is dead. Its several hundred square miles of flood-plain marshes are gone, and its polluted waters

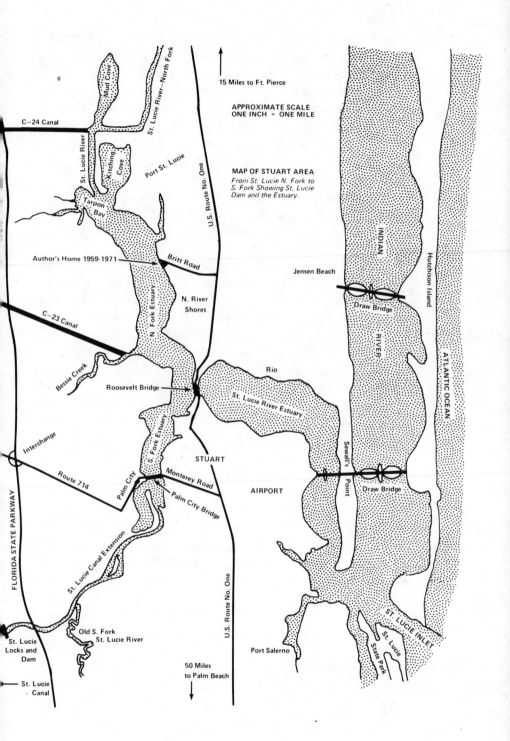

15 Miles to Ft. Pierce

APPROXIMATE SCALE
ONE INCH = ONE MILE

MAP OF STUART AREA
*From St. Lucie N. Fork to
S. Fork Showing St. Lucie
Dam and the Estuary.*

Mud Cove

St. Lucie River–North Fork

C–24 Canal

St. Lucie River

Kitching Cove

Port St. Lucie

U.S. Route No. One

Tarpon Bay

Author's Home 1959-1971

Britt Road

Jensen Beach

Hutchison Island

INDIAN

Draw Bridge

N. River Shores

C–23 Canal

N. Fork Estuary

RIVER

ATLANTIC OCEAN

Bessie Creek

Rio

Roosevelt Bridge

St. Lucie River Estuary

Interchange

S. Fork Estuary

Sewall's Point

Route 714

STUART

Draw Bridge

Palm City

Monterey Road

AIRPORT

Palm City Bridge

FLORIDA STATE PARKWAY

St. Lucie Canal Extension

ST. LUCIE INLET

St. Lucie Locks and Dam

Old S. Fork
St. Lucie River

U.S. Route No. One

St. Lucie State Park

Port Salerno

St. Lucie Canal

50 Miles
to Palm Beach

now rush into Lake Okeechobee through a straight gutter. The pathetic transformation of this once beautiful river was accomplished by the Corps despite strenuous opposition and alternative plans offered by the United States Department of the Interior, the Florida Game and Fresh Water Fish Commission, and the public. The river's demise exemplifies the poor planning, the persistent and deceitful means of pursuit, and the usual disastrous results that have become the hallmarks of the Corps' activities.

The original source of the river was in Lake Kissimmee, some forty-five miles north of Lake Okeechobee. In its downstream course, the Kissimmee meandered for ninety miles through virtually virgin wilderness. It was nationally famous as a prolific producer of largemouth bass. The game in this area ranged from rabbit and raccoon to wild turkey, deer, panther, and wild boar. The adjacent marshes provided breeding and wintering grounds for virtually thousands of shore birds and waterfowl. The Kissimmee region was one of rugged, virginal beauty, and as the Corps so condescendingly states on page 46 of House Document 369, (a 323-page report from the Corps to the Second Session of the 90th Congress, dated July 30, 1968) "the water of the Kissimmee River is of the best quality to be found in southeastern Florida." This was in 1958, before it transformed this national asset into a liability through its penchant for destruction of all things wild.

Over the years, by a series of interlake canal connections, the source of the Kissimmee River was pushed northward another fifty miles to the suburban areas surrounding Orlando. This greatly increased the secondary sewage that seeped into the river from these urban population areas. Also, the heavy increase in agricultural nutrient wastes from enlarged agricultural areas drained into the adjacent streams, and eventually into the river. Despite this several-fold increase in pollution which affected the Kissimmee, its adjacent marshes were able to absorb these materials and transform them into usable components essential to the support of terrestrial and aquatic life. Even with this extra demand placed upon the filtering and absorption capabilities of the marshes, the Corps recognized the purity of the Kissimmee's waters. Yet, with full knowledge of these facts, the Corps obliterated the marshes protecting the river!

In dry periods, the waters of the Kissimmee were confined within the banks of the riverbed. Medium to heavy rains would cause spring overflows, in varying degrees, into the adjacent marsh area. This varying flow of water, from its permanent bed through the littoral zones and into the fluctuation zones, was essential to the river's function as a viable natural laboratory. This water movement that came with the wet seasons might involve areas as much as a mile on each side of the river marshes. It varied, of course, with the elevation of the land. The tree line was usually found at the edge of this configuration. During periods of torrential rains, not an infrequent occurrence in central Florida, the flood waters would extend beyond the tree line—in extraordinary cases, as much as four or five miles.

The tranquility and normal functions of this vast watershed were often interrupted by the expanding cattle interests. In times of flooding, there were often serious losses of cattle, and agitation by the ranchers for flood control continued to grow. The blame was put on the river, not on the fact that the ranchers were intruders on a normal flood plain.

The Corps' first engineering proposals called for cutting a channel 150 to 200 feet wide and some thirty feet deep through the oxbows of the Kissimmee River. The channel, supposedly, would drain all the adjacent marshes and give a runoff of flood waters so that none would reach the tree line and certainly not flood beyond it. The United States Department of the Interior and the Florida Game and Fresh Water Fish commission immediately recognized the absurdity of the Corps' plan. At great length and in scientific detail, they pointed out to the Corps the biological disaster that might occur. They offered a counter proposal—the comparatively simple matter of confining the flood waters to the approximate tree line. Such an arrangement would have been far less costly than the Corps' multi-million dollar proposal. It would have served all the flood-control purposes served by the present system and caused a minimum of disturbance to the natural ecology of the watershed.

The Kissimmee had a remarkable flood-control system of its own. Until 1958, the mean daily flow for the period of record was about 2000 cfs (cubic feet per second). The lowest rate of flow was in May, with 1,000 cfs. The mean high of 3,800 cfs occurred in October. The heaviest flows recorded up *until recent times* were 17,400 cfs on October 6 and 7, 1948, and 20,000 cfs in August 1928. These high flows occurred during periods of torrential hurricane rains. At other times, even with very heavy rainfall, the holding effect of the marshes and flood plains held the flow between 1,000 and 3,800 cfs. There were definite sponging of water and water-storage characteristics to the system. These, among others, were the benefits the Department of the Interior and the State of Florida wished to preserve by stopping the flood waters from spreading beyond the tree line with a series of inexpensive low levees. But this was not to be, for all practicality succumbed to the frenetic actions of the Corps—drainage and dam-building. The damages would be horrendous, and only the land speculators would profit.

These portentous results and this type of skulduggery prompted Dr. Robert M. Ingle, of the Florida Board of Conservation, to state publicly in his October, 1968 interim report on his study of the St. Lucie estuary:

> One situation is becoming increasingly clear throughout this general area. Investors, land speculators, farmers, and cattlemen enter into and develop lands usually considered of marginal utility due to the possibility of floods and inundations. Later when the known peril becomes an actuality, these persons who have established themselves at a recognized risk, beseech public officials to relieve them of their damages even when such relief threatens to ruin already established and substantial business outside the area. (In this case, an established public resource of extraordinary monetary, aesthetic, and recreational value.)
>
> Any relief structures constructed and operated thusly have the element of land reclamation about them even though they may come under the heading of Flood Control Projects. This is even more true when such drainage projects make usable land from wetlands that cannot be utilized until the water has been largely removed.

The Corps' continuous manipulations of the waters in central Florida brought about one obvious and logical conclusion: there was a definite need for a third major outlet from Lake Okeechobee. The practice of impounding flood waters in the lake, now that the runoff into the lake had been enhanced both in volume and speed, posed some very real dangers in case of levee failure. In the event of a torrential hurricane, the impounded waters would become a lethal time-bomb with potential catastrophe beyond comprehension.

Even the Corps conceded this point. In its first response to public demands, it conducted an extensive engineering study concerning the feasibility of a third outlet from the lake. The published report of this survey (Partial Definite Project Report, Central and Southern Project Report for Flood Control and Other Purposes, Corps of Engineers, Office of District Engineer, Jacksonville, Florida, March 28, 1955, Serial No. 19), although stamped "Not For Public Release," made the following points:

> The Overall Problem, Lake Okeechobee is the major water storage and conservation reservoir for the Central and Southern Florida Project. Maximum use of Lake Okeechobee to serve the area *depends on the provision of an adequate levee protection system with sufficient outlet capacity to insure regulation of the lake within safe limits.*
> . . . .
>
> If very high storage levels are permitted for prolonged periods during critical flood years, *the hazards from possible levee failures are increased.* (Italics by the author.)

The wisdom behind these announcements was never tested until the fall of 1969, when a series of meteorological events shook the Corps' theories.

The September rainfall between Orlando and Lake Okeechobee averaged 8.8 inches. Early in October a tropical storm, Jenny, dumped nine inches of rain over a small segment of the Kissimmee River near Sebring, dwindling to as little as three inches in other sections of the watershed. After this brief deluge, rainfall in the area was negligible, but the potential effects of any major storm were becoming increasingly evident. The Kissimmee, because of the channelization, was flowing at the rate of 6,000 cfs as a result of the 8.8 inches of rain in September. Lake Okeechobee had risen slowly and the Corps had to open the flood gates of the dam in the St. Lucie Canal to a flow of several thousand cfs. On October 4, 1969, the *Miami Herald* and the *Stuart News* reported that 21,600 cfs were flowing from the Kissimmee into the lake. The following day the *Herald* reported the flow as 24,000 cfs. The lake stood at a level of 16.14 feet, or 9.24 inches above what is considered to be a safe regulation height for that time of year. According to the Corps, the conservation areas (impoundment areas south of Lake Okeechobee) were full of water. The Everglades were taking all that the limited capacities of the southern sluices would give them. The Caloosahatchee and the St. Lucie were open.

On October 21, the *Stuart News* reported that the lake still remained at the unsafe level of 16.2 feet. The St. Lucie Canal was flowing water out to the sea at the rate of 6,100 cfs, the Caloosahatchee at 4,300 cfs, while the conservation areas were taking 2,500 cfs into Everglades National Park. This meant that 13,000 cfs were being drained from the lake while the intake was approximately 30,000 cfs—24,000 cfs from the Kissimmee and about 6,000 cfs from Fisheating and Taylor Creeks. Even the ingenious manipulations of the Corps' slide rules could not dispute the fact that the lake's volume was increasing by an additional 17,000 cubic feet of water *every second*. This situation was alarming enough, but just some 300 miles away hurricane Laura was churning up the Gulf of Mexico—and Lake Okeechobee lay directly in its predicted path!

Hurricanes are often capricious, and their paths can be predicted with only a limited degree of certainty. Fortunately, Laura veered away from the Florida coast, but the need for another major outlet from Lake Okeechobee was dramatically emphasized.

The logical place for this outlet was along the south shore of the lake. In addition to the matter of safety, proponents of the project were quick to point out other allied advantages:

(1) The already overloaded eastern agricultural canal system would be relieved of further burden and additional expensive structures.

(2) Water levels of the Conservation Areas could be controlled with greater efficiency. Deer and other wildlife would benefit.

(3) By routing the waters that eventually reach the Everglades National Park through the western part of the Conservation Areas, an adequate supply of water would be assured for Big Cypress Swamp. (The National Academy of Sciences has declared the Big Cypress Swamp essential to the continuing viability of the entire South Florida water supply situation.)

(4) A much greater and more consistent water supply would be available for the needs of Everglades National Park.

(5) The Everglades, Conservation Areas, and agricultural districts would benefit from rains limited to the Okeechobee–Kissimmee watershed.

(6) The silting of the St. Lucie Estuary would finally come to an end, and perhaps some measure of life would be restored there.

(7) The spending of millions of dollars for the prevention of salt water intrusion would be unnecessary.

(8) Substantial contributions would be made to supplying the additional sources of water that Congress had been asking the Corps to secure.

(9) The third outlet would transform the whole situation from one of dumping excess water out to sea to one of water management.

The Corps recognized many of these facts in its initial survey. It recommended that a third outlet be built in the form of a mile-wide floodway from the south shore of Lake Okeechobee extending into and through Conservation Area No. 3 with the water routed to Everglades National Park (see map, page 25). Some sixty pages of the report were devoted to engineering analyses, structural designs, and cost estimates—about $15,000,000.

This engineering report was dated March 28, 1955. This was about the time when Col. E. E. Kirkpatrick replaced Col. Schull as District Engineer. Col. Kirkpatrick served until June 18, 1957. Inasmuch as this engineering report did not come to the attention of the public until September 1958, during the tenure of Col. P. D. Troxler, this implies that from its inception until 1958 it was under study by three different District Engineers for a period of about five years. Such a long period of time seemed to indicate that there was considerable agreement in the Corps as to the need for a third outlet. Or was it a deliberate delay to assuage the public's ire?

Regardless, during September 1958, the Corps held a public meeting at the Martin County Courthouse for consideration of a third outlet. The *Stuart News* reported the following account of that meeting:

> With an undertone of possible future disaster, unless the conditions are corrected, U.S. Army Engineers and Flood Control District officials here Tuesday heard warnings that the greatly increased flow makes Lake Okeechobee control impossible unless there are additional outlets.
>
> A formidable delegation of officialdom filled the Martin County Courthouse to overflowing. It was led by Col. Stephen E. Smith and six representatives from the office of the Chief of Engineers at Washington, D.C. Also present were six staff members from the Corps' South Atlantic Division at Atlanta, nine members from the District Office at Jacksonville led by Col. Paul D. Troxler, and representatives of the Central and Southern Florida Flood Control District. FCD notables in attendance were B. Arnold, Executive Director, W. Storch and J. Clawson and several consultants, including E. Frierdman, of Miami, Gordon Gunter, of Ocean Springs, Miss., and J. Johnson of Berkeley, California.
>
> Further in attendance were many county and city officials. . . .
>
> Col. Troxler introduced Mr. E. W. Eden, Chief of Planning for the Jacksonville Office, who said that studies are underway on several floodway and canal proposals

south of the lake. Mr. Eden pointed out the high value of the agreement between the Cattlemen's Association and the local River League in seeking the third outlet.

While east coast Martin County interests were presenting their pleas to the Flood Control officials here Tuesday, Collier County Commissioners of the west coast endorsed a third outlet from the lake with some moves of their own. Their contribution to the cause was the approval of an eighty mile long canal in that area.

Several of the people in attendance at that meeting were to figure prominently in future negotiations. A follow-up news item several days later clearly indicated that the general consensus was that a gentlemen's agreement had been reached and actual construction was simply a matter of putting this understanding into action.

But, as the public was soon to learn, the road between agreement and action by the Corps is often long and devious.

の§を   の§を   の§を

Intoxicated with the "progress" of its march across Florida, the Corps resurrected an old nemesis from the archives of past fallacies: the Cross-Florida Barge Canal. Without a doubt, this was one of the most inane fiascoes ever perpetrated to extract millions of dollars from the American taxpayer's pocketbook.

But one cannot talk about the Cross-Florida Barge Canal in the past tense, for although it has been stopped by Presidential order, the project is still very much alive. The Florida Canal Authority and its political pursuivants are pushing harder than ever to renew construction of this despicable ditch, despite repeated failures to justify it on an economic basis. In this era of high-speed transportation, the Cross-Florida Barge Canal is about as practical as a new horse and buggy trail from Washington to Philadelphia. The distances and economic justifications are comparable.

The planned route of the canal runs from Palatka on the St. Johns River to Yankeetown on the state's west coast. It would flood 27,350 acres of the Oklawaha River Valley—a river, freeflowing, wild, and beautiful—a river shaded by Spanish moss, magnolia, cypress, sweetgum, laurel oak, blue beech, red maple, and loblolly bay—a river that is home for wild turkeys, limpkins, anhingas, rails, bitterns, and herons. This 150 feet wide slash across Florida's midlands would sever the main arteries of the state's natural water flow. The ecology of all mid and southern Florida would be in jeopardy.

The Cross-Florida Barge Canal had a long incubation period, but the Corps was as patient and persistent as an old setting hen. It was determined to see its moronic offspring hatched. It could be a welfare case for generations to come. The original idea for a canal across the state dates to the days when pirates ruled the waters of the Keys and West Indies. It had its first feasibility test during the depression of the 1930's as a "make work" project. Men and mules went to work, but the idea was soon abandoned. It was decided the canal was too expensive and, furthermore, it threatened the water supply of southern Florida. But Congress had committed itself, and even if foolishly so, it was not about to let an empty pork barrel go drifting by. In 1942, enemy submarines were taking their toll of shipping off the Florida coast. This provided the ideal excuse to revise the project again. Congress, with advice from Corps strategists, decided a cross-state canal was the way to outfox the enemy. The Corps was authorized to proceed. When it came up with a benefit-cost ratio of nineteen cents on the dollar, defense officials decided to take another look. In their suddenly enlightened military judgment, they now agreed that ships entering and leaving the canal would be sitting ducks for the enemy. The project was abandoned once again. But the odor of pork lingers; politicians, land speculators, and others who had a selfish interest in the project kept the fire burning.

The Corps was left with an unused authorization. It sharpened pencils and re-juggled books. By the time the war was over it had upped the return to $1.05 on each dollar spent This was still not good enough for such a large investment. A 1958 figure failed to make the grade, but in 1963 the Corps got the green light on the promise of a 1.5 return. This took the most superb bit of hocus-pocus pencil-pushing ever to flow from a piece of graphite and cedar. Construction costs had gone up more than 35% since the Corps' 1945 "guesstimate." And even though this new plan was an expansion of earlier versions, the total cost figure was slashed 13.5%! Of course, this report was padded with such benefits as "land enhancement" and "recreation." Recreation? When the Corps takes one of the most beautiful fishing and boating rivers in America and turns it into a straight barren ditch, what the hell is it talking about?

The results of erroneous calculations and misdrawn lines within the Corps' paper tomes of blueprints now lie indelibly etched upon the Florida landscape. The St. Lucie, the Kissimmee, the Caloosahatchee, the Barge Canal, all stand as sorrowful victims of the Corps' persistent homage to its own tragic miscalculations.

CHAPTER 3

Masters of Deceit

HOW CAN THE Corps continuously maintain such a lofty position—so sacrosanct, so untouchable? Certainly it has not been with public accord for the Corps has been admonished, ridiculed, and threatened, but still its perch is secure. Its actions have outraged citizens and their officials all across the land, including Presidents of the United States, but all have failed to exert any durable degree of control over this unyielding body.

Former Secretary of the Interior, Harold Ickes, had this to say about the Corps: "A no more lawless or irresponsible Federal Group than the Corps of Engineers has ever attempted to operate in the United States, either outside or within the law."

In the July 1969 *Playboy*, Supreme Court Justice William O. Douglas branded the Corps as "Public Enemy No. 1."

વ§ે∾   વ§ે∾   વ§ે∾

An understanding of how the Corps attained such an autonomous position, and the manner in which it operates from this position, will help clarify its tenacious hold on America's waterways.

Congress created the Army Corps of Engineers in 1802 and subsequently charged it with both military and civilian responsibilities. The Corps' power grew until today it is awesome—not militarily, but politically. The star-studded generals and their brassy subordinates have become the most powerful lobbying and pressure group ever to come to Washington.

Congress gives the Corps more than $1½ billion annually for pork-barrel projects. Congressmen are anxious to get their respective shares of this bonanza; it makes a notable impression upon some of their profiteering constituents. And the Corps sees that it does so, for the Corps is quite adroit at parlaying its propaganda through public meetings, local employment opportunities, increased sales of needed materials, and similar facades that will obscure the portent of the planned project.

One may wonder how a simple localized problem—cattle losses on a flood plain that is normally flooded six months of the year—can be developed into an endeavor that will eventually have adverse effects on thousands of lives. Basically, this is how it works. A request is made to the local Congressman—usually by individuals or a group that stand to profit financially—that governmental steps be taken to mitigate a certain situation. The Congressman seeks approval from the appropriate committee (a case of mutual back-scratching) for the Corps to conduct a feasibility survey. Often, economic justification is hard to come by, but the Corps is a master at juggling figures and it is reluctant to pass up an opportunity for self-aggrandizement. As with the Cross-Florida Barge Canal, figures are juggled until the cost-benefit ratio has a favorable ring to congressional ears. Now, if engineering surveys are deemed practical, Congress approves the project and appropriates the money; contracts are awarded and the dirt begins to fly.

This plan of procedure seems simple enough, but it is often long and complicated; it does take time to overcome public disapproval. But the plan is so devised that proponents of a cause have numerous opportunities for appeal and re-evaluation. If the Corps approves a project, it is not about to let it die.

The inseparable relationship between the Corps and Congress continues on an unusually amicable basis because each protects the other. The Corps gives Congress the "honor" or the perogative of initiating the projects, while it sticks to its time-worn philosophy of "we do only what Congress asks us to do." In return, Congress protects the Corps from repeated attempts to limits its authority. In this way, the Corps maintains its immunity, and Congress has a direct pipeline to pour money back home.

The sad part of this wanton relationship is this: our country suffers because each partner is primarily interested in self-perpetuation. The engineer fraternity boys grow stronger, and Congress has found a patsy to help ensure its reelection. To quote an old cliché, "It takes two to tango," and Congress must share some of the blame that is heaped upon the Corps so profusely.

General accusations are easily made, but in Florida, the Corps has spread the evidence for its own conviction. It lies naked in the sunshine for all to see.

＊＊＊ ＊＊＊ ＊＊＊

Enlarging the St. Lucie Canal revealed the Corps' ineptness as engineers. The banks of the canal were cut perpendicularly instead of being sloped, and when the flood gates were opened in the dam, water velocities became so high that loose materials were carried downstream in a torrent. This condition was augmented by another engineering blunder: the flood gates were designed to pass water underneath. This meant that when the flood gates were opened the principal force of the water flow was concentrated along the bottom. Anything not anchored down was washed down the canal and through the flood gates. Sandbars soon began to form a few miles downstream from the dam in the vicinity of Palm City (see map, page 25).

At Palm City, the down-river section that takes the flow from the dam widens out sharply. Consequently, water velocities decrease and the water suddenly drops its load of sand and silt. This soon made boating impossible, smothered vast areas of oyster beds, and ruined the fishing on the previously highly productive flats. The lethal effects of this pollution involved many square miles of water, because the finer silts and colloidals were carried further downstream. The bottom of the estuary was being turned into a biological desert.

Screams of anguish soon reached Congress. On November 15, 1954, the Senate Public Works Committee asked the Corps to investigate. The "accused" was thus being asked to act as Special Investigator for the Court!

In a classic example of its delaying tactics, the Corps took until April 27, 1956, or nearly a year and a half later to render a report. Of course, the Corps is not noted for simplicity when it comes to filing reports; this one took twenty-five closely printed pages of Senate Document No. 6, 85th Congress, 1st Session to announce what was already very obvious. Yet essentially, it was all in the following two paragraphs:

> Paragraph 27: Studies by the District Engineer supplemented by those of the University of Miami, Marine Laboratory, support the contentions of the citizens of the Palm City-Stuart area, that fresh water releases through the St. Lucie Canal are responsible for the deposition of sediment which has caused severe shoaling in the south fork of the St. Lucie River and that changes in salinity caused by such discharges are detrimental to marine life in the Estuary.
>
> Paragraph 28: It appears therefore that local interests have justifiable complaints in this matter. The adverse conditions referred to above, however, appear to be consequential to operation of a canal which is an essential part of a larger water resources project in which there is

Federal and non-Federal interest. This is an unfortunate case wherein carrying out a large improvement, some people are damaged unavoidably in order to achieve a broader public interest.

Thus, in the first paragraph, the Corps admits the cause and result of sedimentation, but in the second paragraph it tries to justify these conditions by such flowery words and phrases as *consequential, unfortunate case, unavoidably,* and *broader public interest*. And, just in case these words were not strong enough to allay public demands, this same Senate Document contained the following statement referring to a letter from the Corps' Major General E. C. Itschner:

> Local interests have assumed as an item of local co-operation the obligation to hold harmless the United States free from damages. He therefore recommends no Federal participation in rectifying the conditions that exist in the vicinity of Palm City.

This statement brought to light another ruse used by the Corps to assure its immunity. The Corps will not enter into any agreement that does not contain this "hold harmless" clause. In this way, it is protected from legal action because of engineering blunders or other questionable procedures. But in the St. Lucie case, there was competent legal opinion expressed that the Corps may have dug a hole in its own protective barriers. Senate Document No. 6 put the Corps on record as being the cause of admitted damages to the estuary. Considering this admission, and the fact that there is a recognized axiom in the basic law of the land that no man can knowingly harm another, there was considered opinion that the Corps' continuance of performance since 1956 could be adjudicated as willful negligence.

Why this legal opinion was never tested is a matter of conjecture. Perhaps it was due to the ironic twist of fate that favored the Corps during its year and a half of investigation. The years 1955 and 1956 were dry, and the flow through the canal was minimal, amounting to a total of only 3,500 acre feet. The rate of siltation lessened, and fishing improved.

These conditions may have lulled the citizens into an apathetic attitude, for all seemed well. Also, the Corps' blatant attitude about legal suits was well known. It is difficult to win anything tangible in a case where the defendant can truthfully boast that it will use *your* money to defend itself and *your* money to settle any judgments.

But this sense of security was short-lived. The trouble started again in 1957 when 800,000 acre feet of fresh water poured through the canal. In 1958, this figure more than doubled with a whopping 1,900,000 acre feet of water being rushed out to sea—all this at a time when the Everglades National Park was frequently pleading for water to replenish its dwindling supply.

Among other things, this bit of irony aroused the public's ire to an even higher degree and renewed demands for a new concept of water management. Coming so soon after the questionings of Congress in Senate Document No. 6, the Corps took immediate steps to soften the effects of these growing complaints. The means by which it hoped to accomplish this feat was the issuance of a five-page booklet on May 12, 1960, titled *Fresh Water in St. Lucie Estuary—Good or Bad?*

Had the Corps wanted to issue a truthful report, its own files contained a wealth of scientific evidence about the estuary. It had already spent a substantial sum of the taxpayers' money for research by the University of Miami. The highly respected James F. Murdock had filed a report on this research in January 1954, for the University's Marine Laboratories. This report was apparently too factual for the Corps' purposes. Murdock reported the truth of sedimentation versus salinity much too well. The Corps completely ignored these findings, which included the following:

> As long as it remains necessary to release water from Lake Okeechobee through the St. Lucie Canal in order to control the level of the lake, there will be difficult problems in connection with boating, fishing, and recreation of the Stuart area.
>
> Temperature, dissolved oxygen, and pH do not appear to be of great importance.

Photo courtesy of Ed Gluckler
10,000 cubic feet per second of silt-laden water rushing through the flood gate of the St. Lucie Canal Dam at Stuart, Florida.

Evidence is presented that severe and rapid changes in the salinity occurred as a result of release of water from the lake. The lowering of the salinity is severe enough to cause a temporary exodus from this area of species of fish preferring a more saline habitat, and could cause the death of species unable to migrate from this environment. Commercial fishes such as pompano, bluefish, mackerel, and trout avoid the fresh water outflow from the canal and during periods of water release, commercial fishing is driven temporarily out of the estuary. Some commercial fishing continues to take place outside the estuary at these times. Sportfishing is practically non-existent in the estuary during the periods of water releases but continues at much lower intensity outside. Sailfishing is carried on well off-shore and is apparently not affected to a marked extent.

Sediments are being deposited throughout the estuary. There are immediate damages to navigation, boating, recreation, and esthetic values. The effects of the sediment being deposited now will remain long after water releases cease. The control of sedimentation is considered the most important problem involved in the water releases.

The salinity changes and the sediment depositions are sufficient to cause substantial damage to the estuary ecology and fisheries of the area. More detailed studies on the permanent effect of sediment on the marine life are needed.

It is recommended that detailed research be conducted on the sediment regimen of the canal-estuary system. The objectives of such a study should be to solve the sediment problem physically and biologically and to establish the possible value of changing the duration and the volume of flow.

Now, most certainly, the Murdock report would not serve to deceive the public into believing that the flows from the St. Lucie Canal were actually beneficial to the estuary rather than killing it; and obviously, it did little to defend the Corps. So another line of attack was attempted.

Dr. Gordon Gunter, a prominent biologist from Ocean Springs, Mississippi, was engaged to make a new biological survey. The Corps sent a biologist from its Jacksonville office to team with Dr. Gunter. Seemingly, this move would produce results more favorable to the Corps. But it would have been most unusual if something so obvious as the silted conditions

These photos of the high vertical banks of the St. Lucie Canal reveal how the wakes of passing boats undermine the banks causing them to collapse. Over the years, these vertical banks have receded as much as fifty feet beyond the original cut. This erosion contributes significantly to the millions of cubic yards of silt that has, and is, choking the life from the St. Lucie Estuary.

could possibly escape the notice of a man of Gunter's experience. Excerpts from his lengthy report follow:

> Lake Okeechobee water released through the St. Lucie Canal carries fine sand, shell fragments and organic material into St. Lucie Estuary. The very fine organic material and clay muck usually suspended in lake waters give it a dark, turbid appearance. When releases are made, the turbid fresh water replaces portions of the water in the estuary. Although most of the fine material is carried into the ocean, some is deposited in places in the bay area where the velocities are very low or in mixing zones of fresh and salt water, which causes the materials to floculate.
>
> The principal source of sand material carried by the St. Lucie Canal is from bank cavings in stretches of the canal between the dam and the lake. The heavier sands picked up along the canal are deposited in the estuary as soon as velocities slow. Hydrographic surveys show the Palm City shoal contained 1,183,000 cubic yards more material in 1954 than in 1932.

Since both these highly reputable scientific sources repudiated the Corps' stand, it decided to issue its own version of how these reports should have been written by publishing *Fresh Water in St. Lucie Estuary—Good or Bad?* This booklet took a few excerpts from the Gunter and Murdock reports; it glossed over the findings of both, and conveniently ignored the real truth about sedimentation.

This report can be evaluated personally by turning to Appendix A where it is carried in its entirety.

చ§§ఎ   చ§§ఎ   చ§§ఎ

During this span of years when the St. Lucie Estuary was being ruined, there was much public agitation for a new approach to reduce the siltations from the canal. It seemed logical that by riprapping the banks, and by creating settling basins to trap the sediments before they washed through the gates, the amount of siltation reaching the estuary could be reduced substantially. The Corps became interested in this proposal, no doubt sensing an opportunity to enlarge the canal. It had long maintained that the St. Lucie Canal was the principal and most effective means of controlling the excess waters from Lake Okeechobee. Mutual interest in the proposal grew to the point where the Corps called a public hearing to evaluate its merits.

Certain citizens of the St. Lucie area were battle-worn veterans when it came to public hearings by the Corps. They were familiar with its *modus operandi*: to create the impression that the hearing is for one purpose, and then after it has begun, to introduce something entirely different, leaving the audience faced with a *fait accompli*. Knowing this, there was considerable correspondence by local citizens with the Corps and concerned members of Congress to establish a clear understanding of the purpose of the hearing. When the public hearing was advertised as required by law, it was announced as being for the consideration of dredging a shoal and a settling basin. The Corps claimed that the shoal posed all sorts of dire dangers to commercial traffic. But considering the situation, there must have been more obscure motives behind the Corps' interest.

When the question of a shoal in the canal first arose, I decided to make some soundings of my own. Since I was not aware of exactly where this shoal was supposed to exist, I decided to sound the twelve miles of the canal from the Indiantown Marina down to the dam. With a friend aboard to assist me, I sounded this area thoroughly. Again, several days before the hearing, I decided to recheck the soundings I had made. The results of both soundings were identical.

At the public hearing, I included the following statements in my report:

"We have made a careful sounding of the St. Lucie Canal from the dam westward for twelve miles to the Indiantown Marina.

"Twenty-eight soundings were made from the safety floats westward to the end of the tie-up pilings. There was a small area in front of the old lock and powerhouse running from 16 feet 6 inches to 15 feet 6 inches. The rest of the area varied from seventeen to nineteen feet. Five more soundings were made from the top of the dam directly in front of the gates for depths of twenty feet.

"For the next twelve miles westward there were three soundings of sixteen feet, and eighteen more soundings ran from seventeen to twenty-two feet. Soundings were made on alternate sides well away from the center line. We used a one-quarter inch line with a three pound lead. Care was taken to assure readings on a vertical line. As further assurance against overmeasurements, the seven-inch length of the sounding lead was ignored.

"We were unable to find any signs of a shoal within or outside of the 1000 foot limitation. Most of the bottom felt hard or rocky to the lead. No extensive signs of soft bottom were found. . . ."

This public hearing, held on May 28, 1969, did not conform to the Corps' usual format. In fact, the Corps virtually ignored the hearing; the Civilian Administrative Chief of the District Engineer's Office was its only representative. In spite of the fact that the Corps has reserved unto itself the full and complete management of the Okeechobee Waterway, of which the St. Lucie Canal is a part, the hearing was chaired by a board member of the Flood Control District. The FCD's Executive Director and the Engineer of Flood Control flanked the chairman.

The opening remarks of the chairman made some vague references to the subject of settling basins. They were so indefinite that various meanings were quite apparent. On the subject of dredging the shoal, the chairman's remarks were very clear and lengthy.

A chart was presented showing certain depths in the waters directly adjacent to the dam. These had been ascertained by the Corps some six months prior to the meeting. There was no doubt that the Corps took the soundings, and that it did find a shoal near the dam involving 20,000 to 25,000 cubic yards of accumulated silt. Also, as a result of this hearing, there was no doubt about the fact that the shoal no longer existed.

Photo courtesy of the Stuart News

This costly Hurricane Protection Structure was built into the Lake Okeechobee levees at Nubbins Slough for the sole purpose of protecting cattle and pasture lands, while several miles to the south, Mayaca Gap remains open and a potential threat to the loss of human life in that area.

The differences of opinion about the mysterious shoal were easily explainable. During the interim of time between the Corps' soundings and the hearing, the Kissimmee watershed received a sudden four-inch rainfall. With the Kissimmee River now channelized, Lake Okeechobee began to rise rapidly. The Corps' first reaction was to open the Caloosahatchee flood gates and then resort to the agricultural canals. These facilities proved inadequate, and it was forced to open the St. Lucie gates on March 19, to a flow of 2,300 cfs. On the following day they were opened to 5,400 cfs and by March 25, the opening carried a flow of 6,700 cfs. The rate then began to taper off reaching a flow of 3,000 cfs by April 8.

Considering the fact that the flood gates open from the bottom, the shoal's disappearance became quite obvious: it was washed under the gates and out into the estuary.

But the Corps, adhering to the strict military discipline that once an order is given it must be obeyed, started dredging the shoal that was not there. No doubt it was an attempt to placate the public's demand for settling basins—at least, the Corps was out there digging.

꿍ᆼᆽ᠍ᐳ   ᠍᐀ᆼᆽᐳ   ᠍᐀ᆼᆽᐳ

The Mayaca Gap is just what its name implies—a gap in the safety levees built around Lake Okeechobee following the 1928 hurricane. It is several hundred feet wide, and it is through this gap that the St. Lucie Canal flows eastward from the lake. The Corps has steadfastly maintained that the gap is specifically for the location of hurricane gates and locks. But it was not until 1972, over four decades later, that soundings for foundations were started on this project.

During this time there was constant concern about the possibility of a hurricane, similar in nature to the one of 1928 that took 2,700 lives, passing over the lake in an easterly or westerly direction. Had this occurred, the destruction and the loss of life could have exceeded that of 1928. Each time a hurricane approached central Florida, the question was, which way will the storm move?

Since the Corps ignored all remonstrances regarding this situation, it became evident that an appeal to some higher authority must be made, although there often seems to be no authority higher than that of the Corps. I asked Congressman Paul Rogers to investigate the matter. He addressed a request for an explanation to the Assistant Director of Civil Works for the Atlantic Division of the Corps of Engineers. The reply to this request is carried in full in Appendix B.

Colonel Edelstein's reply is worthy of some critical evaluation. First of all, it reads like a model of rectitude to the uninitiated. It is embellished with well-known statistics and purposely avoids the real issue as being "extremely remote." All references to hurricanes were for years other than 1926 and 1928 when the tragic loss of lives occurred. The repetition of these conditions was the real point of concern. The reply's reference to Dr. Garbis Keulegan's investigation of hurricane tides needs further clarification; important details are missing. When questions began to arise about the dangers of a hurricane causing considerable damage through the St. Lucie Canal, the Corps quietly began to put itself into a favorable position of apparently having given the matter full consideration. It called in Dr. Keulegan, a reputable scientist, and commissioned him to make a detailed study of the situation and formulate a report on it.

Unfortunately, the Corps did not say to Dr. Keulegan, "The people of Stuart fear that if the forces released in 1928 were to recur in an east-west direction, serious consequences could occur resulting in a great loss of life and property." In effect, what the Corps did say to Dr. Keulegan was, "We want you to take into consideration the conditions of the 1949 hurricane and tell us what would happen if such a hurricane were to recurr."

The Corps issued other ground rules for Dr. Keulegan's investigation. For example, all figures were to be computed on the basis that the flood gates would be closed. Such a situation under hurricane conditions is an engineering joke. Torrential rains precede and accompany a hurricane. Therefore, lake levels would be above prescribed limits, and the gates would be *open* for maximum flow, not *closed* as in the basis of the report.

It would appear that Dr. Keulegan had some misgivings of his own about the sincerity in the request for this report. This is suggested by the repeated references in the report to his instructions from the Corps as per a certain letter of May 23, 1966, in which his limitations to the 1949 hurricane were specifically spelled out. It was almost as if he were serving notice that without these limitations, the report might have been quite different.

Now that the Corps has finally started preliminary sounding work on the hurricane structure in the levee gap at Port Mayaca, one wonders about the real significance behind this sudden spurt of interest in a project that has been delayed for so many years. Surely the Corps has not changed its engineering philosophy about the "extremely remote" dangers of hurricane waters overflowing through the gap. This would be the equivalent of surrender.

If one examines the plans for this new structure, any mystery about the Corps' intent soon becomes quite obvious. It is resorting to an old ruse so frequently employed to force its will upon the people—overbuilding on a project and thereby creating situations that demand additional projects. At Port Mayaca, the new structure will have an overflow capacity greatly exceeding the carrying capacity of the canal. Thus, the stage is set, and it will become increasingly difficult to hold the demands for a larger canal in abeyance. Should this happen, the damages to the St. Lucie Estuary will be multiplied accordingly.

The Corps is also interested in increasing the size of the boat locks from the present 250 feet to 400 feet similiar to the ones it has already built on the Caloosahatchee River. This would be another step toward opening the St. Lucie Canal to increased barge traffic.

꧁ꙮ꧂   ꧁ꙮ꧂   ꧁ꙮ꧂

The Corps, faced with positive scientific evidence that its plan for the Kissimmee River was inviting biological disaster and greatly increasing water control problems for all of southern Florida, still clung tenaciously to its straight-ditch design. It pushed this plan with a frenzy—as if it had a dam-building destiny to fulfill.

Displaying its facade of goodwill to placate the public, the Corps held public hearings, consulted with interested organizations and agencies, and listened to the frantic protests of conservation-oriented opinion. It gave careful attention to the detailed plans of the U.S. Department of the Interior and the Florida Game and Fresh Water Fish Commission, and gave considerable lip service to such statutes as the Fish and Wildlife Coordination Act. But behind it all, the Corps continued to proceed in its own indomitable way.

In a desperation move on December 17, 1958, Mr. F. C. Gillet, Acting Regional Director of the Department of Interior's Fish and Wildlife Service, made a final effort to rescue what he could. Mr. Gillet addressed a five-page statement of position to the Corps in which he included three final proposals. In his letter, Mr. Gillet emphatically states, "The Kissimmee River should be left in its natural state insofar as possible." He called attention to the Corps' previous statement that "there is no means of retaining any portion of the Kissimmee River in its natural state without sacrificing major agricultural benefits desired by local interests."

The "local interests" referred to by the Corps were the cattle interests and a few politicians. There was much pent-up anger in conservation circles that would erupt if given a chance. It was evident that Mr. Gillet knew this, for his letter to the Corps continued as follows:

"We agree that if it is the will of the general public that the Kissimmee River be developed in the manner that you (the Corps) recommend, then the modifications as proposed by the Florida Game and Fresh Water Fish Commission should by all means be incorporated into your detailed project design.

Until that will of the public is established, however, we feel we are obligated to hold firm to our conclusions as documented in the accompanying report."

His letter continued with a firm request for a public hear-

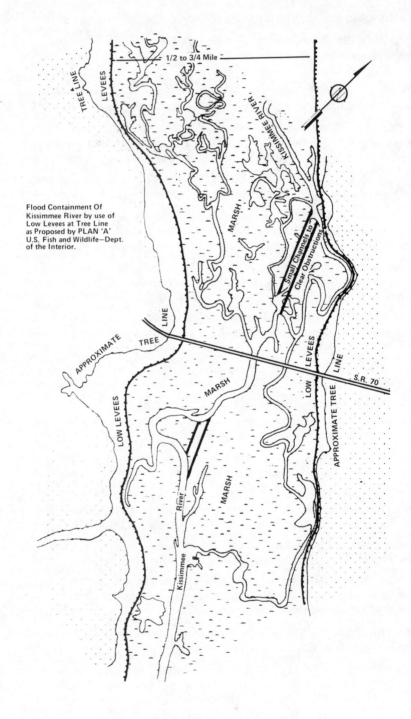

Flood Containment Of
Kissimmee River by use of
Low Levees at Tree Line
as Proposed by PLAN 'A'
U.S. Fish and Wildlife—Dept.
of the Interior.

ing to determine what the general public really wanted. *This hearing was never granted.*

The three plans submitted for the Kissimmee by Mr. Gillet were labeled A, B, and C, but differed only in small detail. A comparison of Mr. Gillet's basic plan and the Corps' plan can be made by referring to the illustration on page 60. One can see easily how the Kissimmee River attained its ninety-mile length by extensive meanderings over a forty-five-mile straight-line distance. One can also see how the levees proposed by the Department of the Interior would preserve some of the natural aspects of the river and much of the marshland.

In the final analysis of the adopted plan, the Corps made one small concession: Florida managed to salvage a few thousand acres of marsh in the lower reaches of the project. But when it was all said and done, the Corps had cut a box channel 150 to 200 feet wide and thirty feet deep right through the heart of the Kissimmee watershed—just as its original plans specified. The Kissimmee was another victim of "The River Killers."

જી⊱   જી⊱   જી⊱

When the great ditch pierced the Kissimmee valley like an arrow on target, the resulting scar flowed with the fluids and elements of life. Water, mud, and sewage surged through the open wound in a vile syrup; Lake Okeechobee was its temporary resting place.

The polluted waters and the system of control structures were items that soon aroused public indignation. Once again the Corps had reached into its old bag of blueprints and installed control gates that opened vertically from the bottom. And once again, the results were the same: when the water flowed underneath the gates, it carried with it all the mud and muck that had settled along the bottom. An alluvial fan spread for miles into the lake. It seems plausible enough that somewhere, somehow, the engineering doctrine of the Corps would have acknowledged the fact that heavier particles do settle in stilled waters, and that water flowing over the top of control gates results in cleaner downstream conditions. The Corps not only refutes this scientific truth, but it refuses to try it, even on the Kissimmee where structures do have skimmer gates on top.

Each time the Corps was approached about this situation, it countered with the usual palaver about "limited flow capacities" and "temporary conditions." Actually, the upper gates have a flow capacity of 10,000 cfs, an amount that is exceeded only in times of flooding. Use of the upper gates during most of the year would do much toward keeping silt out of the lower river and the lake. When necessary, a small dredge could be used to remove the silt from behind the dams.

Although the Corps' procedures are often consistently monotonous, they are sometimes inconsistent. Here, on the Kissimmee, the six dams form natural settling basins, but the Corps refuses to use them as such; on the St. Lucie Canal, it wants to build a number of settling basins to accomplish the very thing it refuses to acknowledge on the Kissimmee.

In the spring of 1969, acting as Chairman of the Pollution Committee for the Martin County Anglers' Club, I wrote a rather pointed letter to the District Engineer of the Corps at Jacksonville requesting that positive action be taken to alleviate the horrible condition of the Kissimmee River. Copies of this letter were forwarded to a number of state and federal agencies including Congressman John D. Dingell, Chairman of the Sub-committee on Fisheries and Wildlife Conservation. It was Mr. Dingell who followed through on the request bringing a reply from the Corps' Chief of Engineers. (See Appendix C.)

❧   ❧   ❧

Even though the reply that Mr. Dingell received was from General William F. Cassidy, it promised nothing about correcting the situation on the Kissimmee. Rather, it was a defensive letter assuring Mr. Dingell that all plans and activities pertaining to the Kissimmee were coordinated with state and federal agencies, and executed "in accordance with existing Laws, Regulations, and Executive Orders." Actually, this is exactly what was *not* done. As previously illustrated, the plans adopted by the Corps were done so in spite of vigorous opposition by both state and federal agencies. The construction and operation of Canal 38 (Kissimmee River) were conducted in a manner that violated and ignored existing laws, regulations, and executive orders.

General Cassidy's reference to conditions "that will provide extensive benefits to anglers, boaters, and sightseers" would be laughable if the actual results were not so pathetic. Where hundreds of boats could be counted on Lake Kissimmee and the river in a single day, now, few are to be found. And how can one speak of aesthetics with any sincerity where a beautiful meandering river has been turned into a straight sluiceway and where the "temporary muddy condition of the river" seemingly has no end?

At the request of the Department of the Interior and the Florida Game and Fresh Water Fish Commission, a couple of feeble attempts were made to maintain some fish habitat along the straightened river. A small shallow canal was dug parallel to the river, with frequent openings between them. This was a failure because the meanders soon became stagnant. Also, the Corps agreed to install berms along the edge of the river so that fish might have a shallow place in which to spawn. Local biologists were never able to locate these berms. Confirmation of these two attempts at habitat restoration can be found in a letter to the author from the Florida Game and Fresh Water Fish Commission, dated August 15, 1969. (See Appendix D.)

Although General Cassidy's letter vouched for the Corps' compliance with legal directives and executive orders, this was the most blatant violation of all.

Executive Order 11288, issued by President Lyndon B. Johnson, directed all federal agencies to "provide leadership in the nation's wide effort to improve water quality through prevention, control and abatement of water pollution from Federal Government Activities in the U. S."

Of course, the Corps was long adept at maintaining the appearance of compliance. As with the Kissimmee project, it would give polite audience to all concerned and then proceed with its own plans.

Executive Order 11288 is quite long and detailed. However, it does contain specific pointed items which make it obvious that the Corps' activities on the Kissimmee were not in compliance with this directive.

Section 1, Item 4: Review and surveillance of all such activities shall be maintained to assure that pollution standards are met on a continuing basis.

Section 1, Item 6: The head of each department, agency and establishment shall ensure compliance with Section II of the Federal Water Pollution Control Act, as amended (33USC,466th) which, as modified by Reorganization Plan No. 2 of 1966 (31 FR 6857), declares it to be the intent of Congress that Federal Departments and Agencies shall, insofar as practicable and consistent with the interests of the United States and with available appropriations, cooperate with the Secretary of the Interior and with State and Interstate agencies and municipalities in preventing water pollution.

Section 3, Item A: The head of each department, agency and establishment shall provide for an examination of all existing facilities and buildings under his jurisdiction in the United States and shall develop, and present to the Director of the Budget, by July, 1, 1966, a phased and orderly plan for installing such improvements as may be needed to prevent water pollution or abate such water pollutions as may exist, with respect to such buildings or facilities.

Section 3, Item B: The head of each department, agency, etc. shall present to the Director of the Budget by July 1, 1967, and by the first of each fiscal year thereafter, an annual report describing the progress of his department, agency, etc., in accomplishing the objectives of its pollution abatement plan.

When the Corps was challenged as to its compliance with this directive, General Cassidy's letter was an emphatic "Yes" that the Corps had complied. However, further investigation was to prove the fallacy of this statement. Initial queries to the Director of the Budget as to whether the Corps had filed the reports as required by Executive Order 11288 brought some rather evasive replies. First, this office states that the Order pertained only to "sewage and general wastes," and that the Corps had complied with this provision. This is an interesting interpretation because the word "sewage" does not occur anywhere in the total content of Executive Order 11288. Actually, what the Corps had done to satisfy this requirement was to install toilets on the barges *Hyde* and *Gerig*.

More specific and persistent inquiries to the Director of the Budget finally brought this reply from Joseph Laitin, Assistant to the Director (November 17, 1970):

"The Executive order you referred to requires that existing Federal facilities meet certain pollution abatement standards. The Corps of Engineers has not prepared a specific report on siltation or water degradation conditions in the Kissimmee and the St. Lucie Rivers and Lake Okeechobee."

Thus, in 1966, the President of the United States said that all federal agencies must abate pollution, must file reports on existing pollution problems with plans for solving them, and once a year must submit reports on their progress. But, by simply saying nothing about the ravages it was committing on the St. Lucie and the Kissimmee, the Corps concealed from the President the conditions of existing pollution which it had been directed to reveal. Since no reports of the existence of the pollution were ever made, it follows that plans for its abatement and annual progress reports were unnecessary.

By keeping silent, the Corps concealed what had been happening on the St. Lucie since 1947–49 and what is also happening there today. It did not file a 1967, 1968, or 1969 report on what progress, if any, it was making there, nor did it make any reports of plans or progress on the Kissimmee. The Corps ignored the Presidential Order.

President Nixon has issued an order similar to Executive Order 11288; in fact, it is very close to being identical. While there may be some minor differences, the basic intent is the same. And, judging from current projects, its insouciant treatment by the Corps is the same.

In the case of the Cross-Florida Barge Canal, it is sufficient to say at this time that the tactics employed by the Corps to impose its will upon the people closely paralleled those of the St. Lucie and Kissimmee projects. However, there was one difference: a new and stubborn obstacle was thrown in their path. The public's concern for the environment had reached such heights that it could no longer be ignored by politicians and elected officials. But even this, for the time being, was just another hurdle to be surmounted by the Corps. The Corps' attitude toward environmentalists is revealed in an article by Tom Herman, a special features writer for the *Wall Street Journal*. In an interview with the Corps, dated January 6, 1970, Herman attributed the following quote to their spokesman: "'Those silly butterfly chasers and self-serving politicians can't stay the way of progress,' snaps one Corps staff official. A Corps spokesman buries his head in his hands and mutters softly, 'Those ignorant, misguided, conceited fools, they know not what they say. We are the nation's leading conservationist group because we have conserved the earth by molding it to suit man.'"

As to the butterflies, the Corps' spokesman was mistaken, for they turned out to be hornets instead, and their sting chased the Corps out of the Oklawaha Valley and forced cessation of work on the Cross-Florida Barge Canal. It was just one year and thirteen days later, on January 19, 1971, that President Richard M. Nixon signed an executive order bringing this comedy of errors to its overdue $50,000,000 end.

Although the demise of the Cross-Florida Barge Canal left a sad memorial etched upon our land, it also brought forth a eulogy that the American people will not soon forget: *the Corps can be stopped.*

This unprecedented action by the President was a severe blow to the Corps' absolute freedom in the planning and construction of waterway projects. In his announcement to the public, the President stated: "We must assure that in the future we take not only full but timely account of the environmental impact on such projects—so that instead of merely halting the damage, we prevent it."

This action, alone, was enough to encourage environmentalists, but other legal directives were also beginning to work in their favor. In the late 1960's they had alertly resurrected the Refuse Act of 1899 from long dormant bureaucratic files. Armed with the poignant statement that "it shall not be lawful to throw, discharge or deposit . . . any refuse matter of any kind or description whatever . . . into any navigable water of the United States . . . citizens had successfully challenged other Corps' projects. And then there was the National Environmental Policy Act (NEPA), requiring that the environmental impact of any project be thoroughly investigated prior to any construction. In a little over a year, over a dozen of the Corps' questionable projects were mired in the legal ramifications of potential environmental damage. Undoubtedly, the early 1970's was the most challenging period in the Corps' history.

The Corps reacted to this rebuff like true soldiers: it came back fighting. Its tactics are difficult to evaluate; they may be sincere, as its intensified propaganda campaign would have us believe, or they be just more self-protecting camouflage. Nevertheless, environmental concern has become an essential ingredient of the Corps' future plans.

The Corps' acceptance of this fact, and the effectiveness of these legal directives, are humbly acknowledged in the latest (as of this writing) news release, dated June 19, 1973, from the Department of the Army, Office of the Chief Engineer. The release states:

> Recommendations on one out of every three major studies by the Corps of Engineers since the National Environmental Policy Act (NEPA) was enacted in 1969 have been changed for environmental reasons.
>
> Major General John W. Morris who directs the Corps' national civil works program today released an analysis of the Corps' performance under NEPA.
>
> He stated that environmental assessments have been made for five hundred projects in the construction and design stages, two hundred studies and over one hundred completed projects.
>
> General Morris said the designs of one-third of the

five hundred projects under construction, or about to be constructed, were modified to accommodate environmental considerations.

Similarly, in almost a third of two hundred studies, the alternative finally proposed had been significantly changed during the study to minimize the impact on the environment.

In addition, a total of 103 completed projects were assessed environmentally. In 43% of these operating projects, different operation and maintenance procedures were adopted to help the environment.

General Morris said that NEPA has accelerated progress toward reorienting staff attitudes from concern only with economic considerations to a better balance between economic, environmental, social and other important considerations which are included in the public interest.

"Although the 'reorientation' is not complete," General Morris said, "real improvement is obvious. Guidelines have been prepared and issued, and as we become more aware of other ways to improve them to reorient our considerations in the planning and design phases, newer up-to-date guidelines will be implemented.

"The Corps' goal in our Civil Works mission is to get the public more involved in the decision-making process," General Morris said. "I feel we have come a long way in the last three years and we are gratified with the assistance we are receiving from a number of governmental agencies and environmental groups and individuals. We must continually strive to get all segments of the public involved in the planning process of our projects.

"We must work to get the many 'publics' together where there will be a free exchange of ideas of the pro's and con's, and can iron out the differences during the planning process—this planning process is where the Corps is 'under the gun,' this is where decisions can be made to go ahead with a project or modify it or to abandon it.

"The Corps has served the public in meeting resource needs in the national interest for almost 150 years. We will continue to apply our professional capability to this service in light of current concerns and values—the Corps cares."

If these statements truly reflect the attitude of the Corps, then there is some vestige of hope that thorough environmental consideration will be given to all future projects so long as we must tolerate its presence. But history repeats itself all too frequently where the Corps is concerned, and this is typical of the forced-fed propaganda used in the past to screen the Corps' real purposes. Neither the Corps, nor some members of Congress, is happy about the impediments imposed upon their enduring partnership, and legislation to ease such environmental encumbrances is under way. There may still be hope. Already a few "friends" of the Corps in both the Senate and the House have been voted out of office on conservation issues. The public will no longer be damned. Consequently, more and more Congressional converts are joining those few who have for so long kept this lonely vigil. A strong and continuing public outcry will result in the pork-barrel running dry and the stilling of bulldozers and draglines.

Meanwhile, the Corps keeps digging and the money flow—which resembles water rushing through an open floodgate—keeps coming.

CHAPTER 4

The Tragic Results

THE TRUE SIGNIFICANCE of the Corps' impact upon the Florida environment has yet to be measured in full. But one thing is certain: there is enough positive evidence that the end of a Florida that used to be grows ever closer.

These may seem to be rather foreboding words but remembering Florida's precariously balanced ecology, and considering the disastrous results already imposed on a suffering land, they ring with an air of obvious certainty. There is visible evidence that the Corps has defiled Florida's waterways, and in Florida water is life. When the water is gone, Florida will be gone, also.

▪▪▪ ▪▪▪ ▪▪▪

Whenever possible, the Corps likes to operate on the premise that it is better to concentrate its planning and activities on a singular isolated project. In this way, it works with local people and caters to local Congressmen, thus avoiding the inherent problems of the cumulative effect on complete watersheds or other large areas. The pathetic part of such philosophy is this: while the Corps concentrates on satisfying the demands of a few, it often initiates incomprehensible results for a much larger area. The St. Lucie Estuary is a classic example.

Death came to the St. Lucie by strangulation. The heavier particles from the massive tonnage of silt that flowed through the canal settled in the Palm City area where the river widened. But the lighter silt, and other materials of organic and colloidal natures, spread from one end of the estuary to the other with the changing of the winds and the tides. Bottom deposits would change as much as two feet in depth with a single shift of the wind. The minute flora and fauna that form the base of the estuary's marine pyramid of survival could not endure in the ever-changing environment.

As if the preponderance of silt alone were not enough to strangle the estuary, other factors began contributing to its demise. Long tendrils of green slime began to appear in the water. It continued to thicken and eventually reached from shore to shore. The nutrient-enriched water was in the process of eutrophication.

The Corps took samples of the slime for analysis. The report: "Algae of benign nature. It will soon leave. It is harmless." The Florida Game and Fresh Water Commission also took samples of its own to be analyzed by the laboratories of the Florida Board of Health. Its report was significantly different: "Several strains of eutrophic types of algae, one of which is dangerously toxic both to animals and to man."

The Corps was wrong when it said the algae would leave. It did not. Instead, it died and fouled everything it touched with a black slimy ooze. The stench became unbearable, and waterfront families were forced to flee until the stench abated because the toxic odors were making them ill.

And so it was in the years 1968 and 1969 that the trauma of death settled across the St. Lucie Estuary. Some years before, Dr. O. E. Frye, Director of the Florida Game and Fresh Water Fish Commission, recognized the impending catastrophe and went on record with this statement in House Document 369, Item IV, page 315:

> With respect to the present system of operation, there are violent salinity changes and excessive turbidities which assist in making the Estuary a biological desert, instead of a nursery and feeding area for fish and wildlife resources.

Even during these fatal years, the Corps adhered to its adamant position that the present methods of operation were beneficial to the estuary. But the loss of the estuary was obvious to everyone. Dr. Arthur R. Marshall, Director of the Department of Interior, Fish and Wildlife Bureau at Vero Beach wrote to me on March 12, 1969 and said:

> We have in reports and numerous correspondence with the District (Jacksonville Office of the Corps) made it rather clear that we do not agree that surges of fresh water and the accompanying sand and silts are beneficial to the Estuary. We have not made any special studies on the St. Lucie Estuary, simply because we thought it totally unnecessary to do so. We are quite well aware that many immobile or slow-moving organisms are killed by large discharges of fresh water into a brackish estuary, and similarly it is quite obvious that a thick blanket of sand and mud do smother out bottom organisms.

On April 23, 1969 Dr. Marshall again wrote to me and said:

> We are well aware that the silt and the sand and the high fresh water discharges made in some years from the St. Lucie Canal are definitely detrimental to the estuary. We know that vast areas of the bottom are covered with soupy organic muds.

But despite public and official recognition of the atrocity, the Corps continues to do what they please.

To the casual observer, the impediments to local boating and the depletion of vast fishing areas are apt to be the most easily recognized damages to the estuary. But in the St. Lucie area, sport and commercial fishing was a way of life, and the economy of the area was closely related to fishing in one way or another. It was this great asset that lured tourists by the thousands; all local businesses profited. The real estate industry flourished because home owners were anxious to be on or near the waterfront. The total economy reflected the loss of the estuary; tourism slackened and real estate failed to hold competitive values. A way of life from surrounding waters was a thing of the past.

The expiration of the estuary incurred losses far beyond St. Lucie's waters. As a spawning or nursing ground for the majority of Florida's coastal and migratory fish species, its loss reverberated for hundreds of miles along the Atlantic coastline. The total impact of this loss is difficult to evaluate, but it is tragic enough to know that one of the greatest life-support areas along the entire Eastern seaboard no longer exists.

ᣚᣚᣚ   *ᣚᣚᣚ*   *ᣚᣚᣚ*

Some thirty miles north of the St. Lucie Estuary there is a vast wetland area known as the St. Johns Marsh. Indirectly, this marsh is allied with the estuary in that canals C-23 and C-24 empty into the north fork of the St. Lucie River. (See location map, page 25.) The marsh, with proper enclosure levees, could serve as efficient impoundment areas, and canals C-23 and C-24 could carry irrigational waters and serve as control outlets in time of need. Such a plan was initiated by the Corps, but the manner in which it applied the plan was a comedy of errors that left two counties with unsolved water problems, and cattle and citrus interests enriched at public expense.

The canals already existed in a minor stage from original small drainage streams that had been slightly channelized. There were little if any problems about canal routes; the marshes were another story. Here, title had to be acquired. It would seem that acquisition of title to the marshes would have been the first order of business. But not so to the Corps; the canals would be built first and then it would pick up title to the marshlands. The protests of local conservationists were ignored.

Naturally, such a situation does not remain under wraps for long. It quickly became apparent that as soon as the canals became operative, the marshlands would be drained dry until levees were built around the marshes to retain the waters. The cattle and citrus interests did not let such an opportunity lie dormant. The Corps finally found that all the marshlands had been bought and were no longer available. The statements from the Corps about the lack of patriotism of the buyers sounded almost childish and petulant.

The levees were never built, and the marshes remained substantially drained. As one drives along Route 60, which bisects this marsh, one can see well-established citrus groves and great herds of cattle grazing on prime pasture lands, all drained dry by courtesy of the Corps and at the expense of a much-needed water supply for two counties.

The lack of common sense shown here is not dissimilar to that displayed by the Corps in its abandonment of an appropriate third outlet from Lake Okeechobee. The loss of the Big Cypress Swamp area, if it occurs, will be traceable to the same lack of foresight the Corps showed in the St. Johns Marsh project.

⇛§⇢   ⇛§⇢   ⇛§⇢

While the St. Lucie suffered its fate by the slow, torturous process of strangulation, the Kissimmee died quickly.

The channelization of the Kissimmee River was undoubtedly one of the most bizarre blunders ever credited to the tainted history of the Corps. In completing this project, the Corps defied public sentiment, ignored scientific evidence, and jeopardized the future of all southern Florida. The predicted results of its defiant pursuit can be measured in the loss of a vast natural recreation area, thousands of acres of wildlife habitat, a natural settling and filtration basin for waters entering Lake Okeechobee, and "water of the best quality to be found in southeastern Florida." They can also be measured by the beginning of eutrophication in Lake Okeechobee and in the land that suffers from the wanton waste of water. In brief, the main artery of the Kissimmee-Lake Okeechobee-Everglades life-line has been ruined.

The most disastrous possibility to emanate from the loss of the Kissimmee would be the biological destruction of the lake. This point was emphatically stressed by opponents of the plan, but the Corps proceeded in its unyielding way. When the project was nearing completion, Jon Buntz, biologist for the Florida Game and Fresh Water Fish Commission, reported:

> The production area of the river is 95% destroyed. The marshes have acted as filtering systems and removed nutrients. Now, the water will be shot down the box canal and no filtering will occur. In other words, the channelization of the Kissimmee River will accelerate the dying of Lake Okeechobee.

There is no doubt that Lake Okeechobee is being polluted by the filth coming into it from the Kissimmee River. This was authenticated in an investigative report dated August 3, 1970:

> To: Boyd F. Joyner, Water Resources Division, Ocala, Fla.
> From: Phillip E. Greeson, Water Resources Division, Albany, N.Y.
> Subject: Lake Okeechobee Phytoplankton Determinations.
> The plankton samples that you collected from Lake Okeechobee during the morning of July 16, 1970 were received in excellent condition. I have examined the samples microscopically and the results are shown on the attached table.
> The phytoplankton population of Lake Okeechobee has changed considerably since your studies began in early 1969. During your first year of study the concentrations of algae never exceeded 4600 cells per milliliter and the concentrations were usually less than fifty cells per milliliter. During that period, *Pediastrum simplex* was characteristic of the algae organisms. It is typical of the early stages of eutrophication.
> The average concentration of phytoplankton in the lake during your most recent visit was 32,300 cells per milliliter. The highest concentration was 106,800 cells per milliliter and was found in the sample from Station 5 located in the western part of the lake. Stations 9 and 12 had concentrations of 66,600 and 12,500 cells per milli-

liter, respectively. Only the sample from Station 2 [located in the southern part of the lake] showed the low concentration typical of past samples. It was sixty cells per milliliter.

An equally alarming situation in the lake is the change of dominant organisms. *Aphanizomenon Holsaticum* was the most numerous algae of all stations except 2. It numbered 95,800 cells per milliliter at Station 5 and 51,900 at Station 9. *Anabaena flos aqua* had relatively large concentrations at Stations 9 and 12. *Anabaena circinalis* was found for the first time. *Morismopedia elegans* showed a tremendous increase in number. It was found at all stations except 2 and 15. At Station 9, off Taylor Creek, it numbered 11,800 cells per milliliter.

Boyd, the increased concentrations of phytoplankton to bloom levels and the change of the dominant organisms indicate rapid acceleration of eutrophication in Lake Okeechobee. The lake, as shown by the characteristics of its phytoplankton, can now be doubtlessly classified as a eutrophic lake.

All dominant organisms were blue-green. *Aphanizomenon holsaticum* is perhaps the most notorious algae for causing lake deterioration. The occurence of this species is so consistently related to hard water that it may be used as an index organism for high pH and usually a high nitrogen and carbonate content. It is rarely found in large numbers except in eutrophic lakes or in polluted slow moving streams. The plant is a frequent component of algae blooms and in favorable habitats it may become super abundant. The plants being buoyant will float to the surface where they form sticky masses. Either alone or in accompaniment with *Microcystis aeruginosa* and several species of *Anabaena,* this plant is frequently responsible for oxygen depletion and subsequent loss of fish.

The report continued with an analysis of a eutrophic lake index and then ended with this ray of hope:

One encouraging sign is that all the *Aphabizemenon* observed were small and emaciated. This perhaps indicates that nutrient conditions in the lake are highly favorable for its reproduction yet unfavorable for optimum growth.

With injurious biological changes now becoming apparent in Lake Okeechobee, it is possible to see how all of southern Florida is in a most calamitous position. The structuring and planned use of the lake is so designed that in time of need, or in the time of excessive flooding, waters can be diverted into the Conservation Areas and on through the Everglades to the National Park. This means that the nutrients, pesticides, and other contaminants that are building up in the lake will be flushed through the Everglades. If the shallow, slow-moving waters of the Everglades become saturated with these contaminants, the ultimate results could defy human comprehension.

The initial results are easier to predict. An abundance of algae blooms would smother submerged vegetation; above-water species would then proliferate. Transpiration would increase, water flow would be slowed, and drought conditions accentuated in certain areas. Wildlife species would change in accordance with the vegetation. Beyond this would come the ramifications of water table levels, salt water intrusion, and similar problems not yet determined.

To lose the Everglades would be to lose southern Florida as we know it today. This is the ominous bequeath we have inherited from the Corps—potentially the most tragic in the Corps' invincible reign.

The radical changes on the Kissimmee and their effects upon Lake Okeechobee can be understood by examining some comparative flow figures. As previously indicated, the mean daily flow prior to 1958 was about 2000 cfs. The torrential hurricanes of 1928 and 1948 caused brief maximum flows of 20,000 cfs and 17,000 cfs, respectively. By examining the accompanying flow chart for the peak month of October (1969), it can be noted that these figures are exceeded now from *normal* daily rainfalls. It is also obvious that the undisturbed marshes and flood plains did an excellent job of slowing and absorbing flood waters.

The following figures received from the U. S. Department of Interior Biological Survey, Water Resources Division, are the official figures for the daily discharges of the Kissimmee

River (Discharge Structure S 65 E) into Lake Okeechobee.
The rainfall figures are official from the Flood Control District.

Kissimmee River

Discharge	Rain	Date
cfs.	inches.	Oct., 1969
6,600	6.11	1
11,200	2.76	2
23,500	2.46	3
20,200	.01	4
21,700	2.9	5
21,900	.12	6
19,300	.3	7
11,600	.3	8
10,600	—	9
9,490	—	10
8,900	—	11
8,680	—	12
8,250	—	13
8,060	—	14
7,930	—	15
7,480	—	16
7,660	1.05	17
8,300	.14	18
8,320	.43	19
7,720	.81	20
7,370	.05	21
9,860	.10	22
8,970	.43	23
6,810	.61	24
6,810	.37	25
6,070	.21	26
6,180	.70	27
5,780	.23	28
4,850	.52	29
4,120	.50	30
4,730	.55	31

FCD Photo

Only a medium-heavy rainfall caused this unbelievable flow of water in the Kissimmee. It is coming through the gates at 23,500 cubic feet per second. In comparison, the same kind of rainfall caused a flow of approximately 3,000 cfs in the Kissimmee *before* the Corps channelized it. (See water discharge data on page 78).

The baneful results of the Corps' indiscriminate planning and procedures continued to be compounded—on Lake Okeechobee, the Kissimmee, the St. Lucie, and elsewhere in Florida and across the nation. Public concern for the environment brought about legislation requiring the Corps and other agencies to prepare and submit environmental impact statements to the Department of Interior on areas of current and future activities.

On May 5, 1971, the Corps' District Office in Jacksonville complied by submitting its *Draft Environmental Statement, Central and Southern Florida Project.* This 118-page document was so evasive and full of generalities that it was soon challenged by the Department of Interior in an analytical reply.

UNITED STATES DEPARTMENT of the INTERIOR
Office of the Secretary
Washington D.C.

15 October 1971

Dear Colonel Fullerton:

The Department of the Interior has reviewed the draft environmental impact statement on the Central and Southern Florida Project dated May 1971.

The Kissimmee-Okeechobee-Everglades area is an intricate ecosystem dependent on life-giving water. In an effort to proclaim dominion over this vast ecosystem, man developed a system of canals and levees justified by economically oriented benefit-cost analysis, with little regard to either the certain and inflexible laws of the natural world or the multitude of non-economic values. The simplified "technological quick fix" cannot be designed to manage the resources of a complex ecosystem, and numerous short circuits are appearing in present project design—water—stressed deer—drought—fire—eutrophication. The biggest failure of the impact statement as submitted is that it fails to treat the project as a series of highly integrated subecosystems. It does not recognize that an action in one part of the system will yield a reaction in another part. Until totally related nature of the South Florida ecosystem is recognized, resource management policies and planning are doomed to failure. Much of the planning for the Central Southern Florida Project was done during a period when the impact of engineering work on the environment received little consideration. The statement is replete with evidences of the philosophy that project plans have been formulated and should not be changed. This is no longer acceptable.

Section 102 of the National Environmental Policy Act of 1969 states: "All agencies of the Federal Government shall utilize a systematic, interdisciplinary approach which will insure the integrated use of the natural and social sciences and the environmental design arts in planning and in decision making which may have an impact on man's environment." The objective of an environmental impact statement is to appraise the effects of a proposal on the environment and to compare this appraisal with that of alternative actions. In the case of a partially completed project such as this, we believe that the already constructed portions should be reviewed in the perspective of time and operational experience to determine which segments should be modified to ease their environmental impact, while at the same time better meeting project objectives.

The statement indicated "this present statement is based on information obtained through past coordination with State and Federal agencies and through studies and inventories by the District staff. Some of those studies are of survey scope, others were carried out in conjunction with the preparation of general design memorandums some years ago. Subsequent developments, including the increased national concern for man's environment, would necessitate consideration of additional alternatives and would increase the probability that the final plan might vary considerably from that previously adopted."

We do not agree that the present statement is based on past coordination because the U. S. Department of the Interior has offered numerous environmental alternatives since 1958 and no significant project alteration has occurred yet. Although twenty-three years have passed since the original authorization, the philosophy of exploiting the resources to the fullest extent possible for economic development while the remaining natural ecosystems "share the adversity" has continued throughout that period and is reiterated in the impact statement. The impact statement has been written more in the contex of an economic development plan which justifies further project works rather than factually reporting the environmental effects of the works.

As indicated on page 22 of the statement "The coastal strip is rapidly approaching its population saturation point."

This is placing all of the ecosystems of South Florida under stress. The stress is indicated by the numerous fish and wildlife species which are on the Department of the Interior's rare and endangered species list. Instead of introducing additional construction modifications which can at best extend the life of the ecosystem a few years, we would suggest easing the stress by reappraising the water resource base under low water conditions instead of average levels, flood plain zoning, water level fluctuation, more modern land-use philosophies, coastal zoning, recycling of sewage water, aquifer recharge, and a general deceleration of growth in South Florida.

In numerous places throughout the impact area statement, the argument is advanced that unless the Federal project is implemented, haphazard private drainage would be carried out by local interests with extensive detrimental results. This argument ignores the present day regulatory powers of the State, potential for adequate flood plain zoning and present day attitudes toward the environment. Comparisons with earlier drainage operations by the Everglades Drainage District are not valid. Ultimate development of the South Florida area should not be considered inevitable since the resource base cannot support it and still sustain quality environment.

The environmental impact statement should be completely rewritten and, as thoroughly as possible, should evaluate the true environmental impact of the project. In the redrafting of this statement, alternatives to the presently authorized work should receive complete evaluation and discussion. The statement should include restoration alternatives as well as nonstructural alternatives.

The following general comments point out the inadequacy of the impact statement in describing the environmental effects of this project. Most of the discussion and alternatives are based on previous correspondence and coordination offered by our agency but which have had little impact on project modifications in the past. Most of the points are not mentioned or are inadequately discussed in the impact statement.

The statement (Corps' Environmental Impact statement) does not list most past possible alternatives. Those which were listed were generally dismissed or disregarded because of economics, engineering design or for some other reason. These alternatives should be pointed out and discussed in the rewritten statement. The history of the Kissimmee River is one such example where greater consideration should have been given to the possible alternatives that were suggested by the Department of the Interior. (This refers directly to the rebuff by the Corps to the suggestions of Mr. Gillet as described in the previous chapter.)

The Fish and Wildlife Service demonstrated in its 1958 report that the river furnished an unusually valuable bass fishery which would be lost if the flood control plan was carried out. Similarly, the value of the wildlife resources would be greatly reduced. In view of the very large expected losses, the Service proposed the consideration of three plans for preserving a portion of the valuable marsh in the Lake Kissimmee to Lake Okeechobee section of the project area. This region has suffered severe loss because the potential for mitigation of such losses was ignored. The intent of the State to restore portions of the Kissimmee flood plain indicates that the FCD now shares the philosophy underlying the original alternatives recommended by the Fish and Wildlife Service.

The original meandering course of the Kissimmee River was ninety miles. After the construction of the Canal, C-38, the distance is fifty-two miles, and the average flood plain area has been reduced by 65%. After authorization, minor modifications consisting of making permanent pools above the five structures were placed in Canal 38. Due to contour errors, the areas which were to be inundated for mitigation (approximately 11,500 acres) were hardly even wet."

This critique on the Corps' impact statement was signed by W. W. Lyons, Deputy Assistant Secretary of Interior. Perhaps its most significant point is that the Corps cannot continue to bluff its way around environmental considerations. The Corps was told in no uncertain terms that the statement was not acceptable and that another one should be prepared according to the guidelines given. Reportedly, it is doing this, only this time it is working in conjunction with local and other federal agencies. Hopefully, this will prevent future blunders such as those perpetrated on the St. Lucie and the Kissimmee.

⋙§⋘   ⋙§⋘   ⋙§⋘

Considering the persistent and defiant attitude of the Corps in all past procedures, it seems inconceivable that it would ever condescend to admit that it may have made a mistake. But such seems to be the case with the Kissimmee River.

On June 3, 1971, Nathaniel P. Reed, Assistant Secretary of Interior, told a House Sub-committee that the Corps of Engineers was considering modification of the $24.5 million project that converted the Kissimmee into a straight ditch. Also, before a meeting of the Florida State Cabinet on December 12, 1972, according to a number of press releases, the Corps agreed that the waters of the Kissimmee had to be cleaned. This meeting was one of several held by the Florida Cabinet to determine how the state might proceed to bring about a reasonable degree of restoration of the Kissimmee flood plain and marshes. Also, these meetings were an effort to clarify the legal position of the FCD. The state was determined to proceed on its own, if necessary, to salvage what it could from this environmental catastrophe. The FCD had developed plans and experimented with certain structural changes which proved that a sizeable percentage of the watershed could be restored effectively.

Testifying before the Cabinet, Arthur Marshall, then a member of the FCD governing board and Director of Applied Ecology for the University of Miami's Urban Studies, stated that the restoration of the Kissimmee ". . . will not be easy, but it is something we must do, or we will be facing a water quality crisis in Lake Okeechobee."

There is a degree of urgency about getting the restoration project underway, but in total concept, it is an untested engineering problem that will take intensive planning to prevent the millions of tons of fill and other debris from washing into the lake. If the waters of the Kissimmee can be cleaned and the quality of water in Lake Okeechobee maintained, then a third outlet from Lake Okeechobee is still a feasible project.

و§ܒ   و§ܒ   و§ܒ

A reign of silence had settled over the issue of a third outlet for Lake Okeechobee. Robert Bair, President of the St. Lucie-Indian River Restoration League in 1958, and one of the strongest proponents of a third outlet, continue to pursue the issue. In October 1958, he received a letter from Colonel Paul

D. Troxler, the Corps' District Engineer, which literally sealed the doom of a third outlet. The "gentlemen's agreement" that had been reached at the Stuart hearing was now being voided. A totally new concept was outlined in a dictatorial letter having a take-it-or-leave-it attitude:

U.S. Army District Engineer, Jacksonville
Corps of Engineers
575 Riverside Ave.
Jacksonville 2, Florida

Oct. 15, 1958

File SAKGW
Mr. Robert T. Bair, President
St. Lucie-Indian River Restoration League
Box 496
Jensen Beach, Fla.
Dear Mr. Bair:

Reference is made to your letter of 27 September concerning provisions of a floodway outlet from Lake Okeechobee. The resolution of the Florida Wildlife Federation on the same subject has also been noted.

Several of the points mentioned in your letter were covered in the recent conference held in this office which was attended by you and Capt. Leighton. I shall summarize the matter as it now appears.

As now envisioned, additional storage of floodwaters would be provided in Lake Okeechobee in two stages, each involving progressively higher lake regulation levels and greater storage increments. The first stage would be made possible, while maintaining adequate safety for the lake area, by raising existing levees, providing levees for the now unleveed northwest and northeast shore areas, and increasing the outlet capacity of Caloosahatchee River. Thus, by providing greater storage capacity in the lake, much of the excess floodwaters could be held in storage rather than immediately discharged at high rates as now often has to be done. The revised plan of regulation and the new works of the first stage would also mean that when discharge was required by way of the St. Lucie Canal, it would be made, insofar as is possible, at a comparatively low rate (about 3500 or 4000 cubic feet per second). This would eliminate the canal erosion problems and sand bar deposition in the estuary.

As discussed at the recent conference, our biologists have found that medium discharges, as described above, would enhance the fertility and nurture properties of the estuary while being made, thus providing more small food fishes for the larger species. The larger species would move farther down river during discharge, but would return after its cessation to feed on the smaller varieties. Also, medium discharges would move the salt water interface downstream, but would still provide good boat fishing in the lower part of the estuary.

It is believed that the ability to store larger volumes of floodwater in the lake and to make releases at medium rates would provide a substantial amount of relief for the Stuart area.

The second stage of increased storage in Lake Okeechobee would be effected when demand for additional storage in the lake becomes apparent to supply the needs of the area. The second stage would require no further levee construction, but would require an appreciable increase in outlet capacity.

The works described above for the first stage would in general fulfill the scope of the Flood Control Project as now authorized. The increased outlet capacity associated with the second stage, particularly if a new canal or floodway to the south is recommended, would in all probability have to be added to the project by the survey report procedure. This would require submission to Congress for authorization.

As you know, however, this office has already made preliminary studies of several floodway routes and other means of increasing lake outlet capacity. This material will be available when the necessity for more detailed second stage planning becomes apparent.

I trust that the above comments will summarize the present status of regulation planning for Lake Okeechobee.

Sincerely yours,
Paul D. Troxler,
Colonel, Corps of Engineers,
District Engineer.

A brief analysis of this letter will reveal its fallacious content. It had already been proven that the capacity of the Caloosahatchee could do little more than carry the excess runoff from its own watershed. Enlarging it would submit the Ft. Myers estuary to the same problems found in the St. Lucie.

And it had been proven beyond doubt (in the St. Lucie) that fresh-water discharges into an estuary do not enhance its properties and fishing values.

Increasing the height of the levees would not "provide a substantial amount of relief for the Stuart area." On the contrary, increased flows through the St. Lucie Canal would be needed in time of emergency as this is still the main outlet for the lake's excess waters. But the Corps knows this, and the "alternative means" referred to in the letter suggests a camouflage in order to double the size of the St. Lucie Canal; the plans are already formulated.

The late Captain Bruce G. Leighton, U. S. Navy, Retired, was also a strong proponent of a third outlet, and he had done extensive research to substantiate his position. Robert Bair sought his advice regarding the situation as revealed in Colonel Troxler's letter. Captain Leighton addressed a critical analysis to Bair, including numerous predictions. Many of these predictions have come to pass and are now a matter of record. The analysis is quite detailed, but it is a most logical and interesting review of the problem. (See Appendix E.)

The Corps' intent soon became evident. What it had considered essential in the form of "other alternative means of increasing outlet capacity" was now quietly abandoned. Increased storage capacity in the lake was the new order of the day, and with little regard for safety. This was accomplished by substantially heightening the levees. Storage levels have now been increased to 16.4 feet above sea level. Official plans of the Corps call for raising this to nineteen feet and later to twenty-one feet. Again, one cannot help but think of the provisions on page 3, Item 6 of the 1955 Engineering Study: "If the very high storage levels are permitted for prolonged periods during critical flood years, *the hazards from possible levee failure are increased.*" (Italics by the author.)

This is where the proposed project stands today. Inquiries about the third outlet bring the reply that it was proved unfeasible and prohibitively costly. Since when does the Corps consider $15,000,000 a prohibitive figure? It is interesting to note from the Corps' own figures, which they have officially filed with Congress in Senate Document No. 6, that the damages they have admitted doing in the Stuart area as a result of siltations from the St. Lucie Canal now amount to more than $20,000,000. This condition would have been corrected by a third outlet—and it would have paid for itself.

The sudden abandonment of plans for a third outlet remains as one of those mysteries that so often surround the Corps' decisions. The Corps had favored the project for five or more years, and the engineering data had been completed. The Flood Control District had endorsed it, and it was strongly supported by the public. Yet, despite these facts, the third outlet was declared dead by Colonel Troxler's infamous letter.

Surely, this death sentence did not originate with Colonel Troxler, because he and his staff promoted it vigorously at the 1958 public hearing. Colonel S. E. Smith and his staff of six officers from Washington also expressed their support at the hearing. And the Regional Office in Atlanta was not likely to issue an order contrary to the apparent desires of the Washington command. Thus, through the process of elimination, it became rather obvious that the decision had to come from, or near, the top echelon of the Corps. This decision, and the immediate secrecy imposed upon it, suggests that the third outlet suffered a quick political death.

If the loss of human life, or some other catastrophe, makes it necessary to resurrect the plans for a third outlet, the Corps will finally have to condescend to the need for regional planning. The enriched and contaminated waters of the Kissimmee cannot be flushed into the Everglades lest the potential catastrophe becomes manifold.

A brief summation of what has happened to Lake Okeechobee makes this a very logical and alarming question.

For years the Corps managed to keep its grip on central and southern Florida by singing the same old theme song: "Danger on the Lake," but the tune was rarely the same. First, it claimed that high storage levels involved possible dangers, and that additional outlets from the lake were necessary. Then the tune became a paradox; the Corps now said that higher levees were the answer and more outlets were not needed. The next move was to raise the levees and enlarge the Caloosahatchee; the latter proved ineffective because the extra capacity was pre-empted by local drainage. In the meantime, a third outlet from the lake was being promoted, but this tune ended with one quick downstroke of the baton.

The lake's latent dangers were multiplied by the channelization of the Kissimmee. The flow into the lake was increased in both volume and speed, but the outlet capacity had not been changed. In addition, the lake was subjected to progressive eutrophication.

The current theme is to build the levees higher and higher, or in other words, to increase the size of the "bomb." Considering these facts, it is only reasonable to ask: "Has the Corps lost control of Lake Okeechobee?"

⋙§⋘   ⋙§⋘   ⋙§⋘

Considering what has already been said about the Cross-Florida Barge Canal, it would be somewhat redundant to justify its demise at this point. However, it should be acknowledged here that a group of devoted citizens, with logic and persistence, can exert great influence toward the protection of America's environment—even stop the Corps!

The Defenders of the Florida Environment (DFE) was such a group. This organization, under the leadership of William M. Partington, its president and a former staff member of the Florida Audubon Society, Dr. and Mrs. Archie Carr, David S. Anthony, Paul E. Roberts, and Martin Mifflin, was largely instrumental in bringing forth the Presidential Order that stopped the canal. Undoubtedly, it was these energetic people to whom General William F. Cassidy referred as "little old ladies in tennis shoes." This phrase has been bantered about by all facets of the news media and may have aided in bringing about a sympathetic public response to this group's endeavors. One thing is certain: when the Corps got kicked out of the

Oklawaha Valley, General Cassidy found out that these particular "tennis shoes" were tipped with steel.

The DFE embarked on a total inventory of all aspects of
the canal and what effect it would have on the environment.
Its findings were published as *Environmental Impact of the
Cross-Florida Barge Canal on the Regional Ecosystem.* Undoubtedly, this was the most logical and forceful analysis of
the entire canal problem. Armed with the information, the
DFW joined forces with the Environmental Defense Fund and
brought suit against the Corps. This suit was pending when
President Nixon issued his unprecedented order.

The Cross-Florida Barge Canal is referred to as being
dead, but actually, it is still very much alive. As these words
are being written, the Florida Canal Authority is in federal
court challenging the President's authority to stop work on the
canal. At least three other suits are pending. If these pending
legal decisions should favor resumption of the canal project,
there are others ready to demand a re-evaluation of its economic justification. It seems logical to assume that these legal
battles may continue for years.

◦§◦   ◦§◦   ◦§◦

There is a very poignant lesson to be learned from the
despoliation of the St. Lucie and the Kissimmee, but it is obviously beyond comprehension by the minds of the Corps. Its
attention is now focused on Florida's largest river—the meandering, wooded Apalachicola—that winds its way from the
Alabama and Georgia borders southward through the Florida
panhandle into the Gulf of Mexico. Under pressure from upstream industrialists, the Corps has devised plans (with Congressional approval) to apply its old formula—dams and
straight-line ditches—to the Apalachicola. (See map, page 91).

The Corps' proposal involves more than $200,000,000 on
expenditures for dams, locks, ditches, and man-made canals
that would transform this beautiful wild river into another
barge route. Florida critics maintain that the following environmental damages would be incurred by the adoption of this
plan:

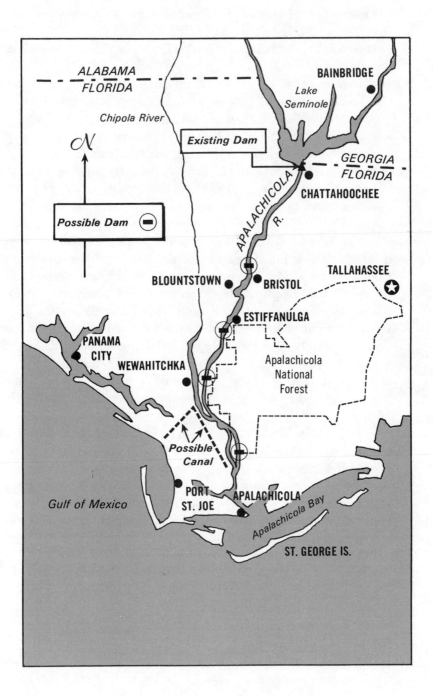

- Destruction of thousands of acres of forest lands—drowned in the artificial lakes created by the new dams.
- Increased hazards from spilled oil, chemicals, and pesticides transported by the increased barge traffic.
- Transformation of scores of miles of natural stream, now abundant with fish and wildlife, into an unsightly "ditch" inhabited chiefly by barges and tugboats.
- And the potentially "catastrophic" chance that damming of the river could ruin the thriving seafood industry that now flourishes in Apalachicola Bay—where the marine life depends for its very existence on the muddy, nutrient-laden water coursing down the river.

The latter point is of the utmost concern, for the bay now produces 90% of the state's marketable oysters. The ecology of the bay is precarious, depending on a balance of fresh and salt water. If the supply of fresh water is not constantly maintained, the oyster's most destructive predator, the oyster drill, would move in with the salt water and destroy the oyster beds.

A researcher for the state's Coastal Coordinating Council, Dr. James Jones, estimates that up to 90% of the shrimp and fish caught in the 20,000-square-mile area offshore from the bay use it and the river as a nursery sometime in their life-cycle.

The Apalachicola has other significant ecological and historical features. Its borders abound with birds and wild game. It, and the tributary Chipola, is the home of the rare Barbour's map turtle. The Chipola is also the only known habitat for a unique species of shoal bass. Both shores of the Apalachicola contain numerous unexcavated Indian mounds. These features would be destroyed if the Corps' plan is adopted.

The sudden interest in damming and channelizing the Apalachicola originated with First District Congressman Robert L. F. Sikes. His "partnership" with the Corps extends over a period of thirty years. As Chairman of the House Appropriations Committee on Military Construction, Sikes has had little trouble in getting what he wants. His district is dotted with pork-barrel installations. However, this time Congressman Sikes is running into some avid and coordinated opposition.

Sensing this public resentment, the Corps is armed with a number of alternative procedures including:

- Construction of four locks and dams on the Apalachicola—a possibility that the engineers have concluded would flood vast areas along the river and would be both economically and environmentally undesirable.
- Build just two dams on the upper half of the river and continue dredging on the lower half—a move the Corps says would still destroy forty-five miles of natural stream and flood 5,000 acres of land.
- Build two dams and channelize the lower Chipola—a tributary of the Apalachicola—which in addition to the damage caused by the two dams would destroy eighteen more miles of natural stream and would eliminate 20,000 acres of marsh and forest lands.
- Build the two dams upriver and leave the lower Apalachicola untouched—bypassing it entirely with a new, twelve-mile barge canal across sparsely populated Gulf County to link up with the Gulf Intracoastal Waterway.
- A fifth plan that would involve increased dredging and diking—probably what the Corps wanted in the first place.

In response to the Corps' plans, Florida's Governor Reubin Askew has suggested that the Corps might do well to leave the Apalachicola alone. The state's Attorney General, Robert Shevin, has threatened legal action if the Corps makes the slightest move to launch one of its plans. And Randolph Hodges, Director of the Department of Natural Resources, has warned the Corps that he is "vigorously and unalterably opposed" to damming the river. In addition, individual citizens and organized groups are largely unanimous in voicing their opposition to any further interference with the river. If Congressman Sikes persists in his sponsorship of the Corps' plan, it might well be that his thirty-year "partnership" with the Corps will be dissolved. By its own admission, any of the Corps' proposals would do irreparable damage to the Apalachicola watershed. The time to stop this mania is now—not after surveying the devastating results.

PART II
THE SCENE SHIFTS

CHAPTER 5

Eastern No-Goods

THE CORPS IS everywhere. The roar of its monstrous machines echoes throughout valley after valley, from the Atlantic to the Pacific. Wherever it can find a ripple on the water, the Corps is ready to "improve" the situation with a dam or canal at the slightest political motivation. Its damn-dam foolishness, either in construction or planning stages, has reached virtually every major waterway in the United States. In the East, rivers like the Potomac, the Delaware, and the Connecticut have either felt the destructive swipe of a bulldozer or are under the scrutiny of the Corps' surveying transits. Armed with some $15 billion in backlog appropriations, the Corps must keep on the move. This pork-barrel money must find its way to the political instigators, lest the Corps' autonomous authority be imperiled.

The Corps' assault on the nation's water resources did not

start on Lake Okeechobee, or on the Kissimmee, or the St. Lucie; it started on the country's largest river, the Mississippi. In 1924, Congress assigned the Corps the task of removing some debris and sandbars from the Mississippi and Ohio rivers. This assignment opened Congressional eyes to the great potential of using the Corps to please powerful constituents at home. Next came California's Sacramento and San Joaquin Basins, followed by the Sacramento River. This plan worked so well that Congress continued to give the Corps additional responsibilities: flood control, navigation, hydroelectric power, water supply, and recreation. All Congress had to do was approve projects and appropriate the money on a mutual basis. The homeward flow of pork-barrel money was unlimited, and so was the destruction of the natural environment.

During its long tenure under the protective shield of Congress, the Corps' philosophy has developed into a mania for dams and barge canals. The Corps has been likened to a dogcatcher—it cannot stand to see a river "running wild;" it must be impounded. Seemingly, the Corps operates on the theory that every city within the country must have a water route to the sea. And it has been this philosophy, and the Corps' exceptionally compatible relationship with Congress, that. has wrought havoc upon the water resources of the nation.

Almost any politically motivated excuse is enough to spring the Corps into action, regardless of the project's impact upon the environment, the economy, or the nation as a whole. Such actions can only exemplify the mutual understanding of self-perpetuation. There can be no other reason, or why would the Corps promote, and Congress appropriate $514 million, for a canal to link the Tennessee and the Tombigbee rivers to facilitate the shipping of strip-mined coal to Japan? Why should sections of northern Alabama and eastern Tennessee be devastated to provide fuel for Japan when there is an energy crisis at home?

This project will require the digging and impounding a 186 mile stretch of the Tombigbee, the largest of the "natural" rivers remaining in the entire Mobile Basin. It will also require a forty-five mile long lateral canal and another of thirty-nine miles to pierce the ridge separating the two river basins. In addition, hundreds of miles of tributaries will be channelled, and an estimated 258 million´ cubic yards of earth will be moved. Digging and damming of this magnitude cannot be done without serious damage to the environment. A report from the regional office of the Environmental Protection Agency states:

> The channelization, the dams and impoundments, the introduction of Tennessee River water into the Tombigbee River system . . . and the construction operations for completing the waterway will all have a profound and lasting effect on water quality values and the ecology of the entire area through which the waterway passes.

In attempting to justify this project on an economic basis, the Corps reverted to one of its old tricks by presenting only one side of the question. It tried to point out the savings to shippers using the canal without giving proper consideration to existing rail and road facilities that are adequate to do the job. Again, the taxpayer is being asked to subsidize the economic and political interests of a few. Being adept at juggling figures to meet the need, the Corps has managed to double the cost-benefit ratio within the past five years. It did this by altering the rate differential between railroad and barge traffic. This is a flagrant violation of the 1966 Transportation Act, so the whole project is legally questionable.

Although citizen groups have obtained a temporary injunction, considering the factors of environmental damage, unneeded transportation facilities, and illegal procedures, the Congress should show its interest in the country by immediately deauthorizing the Tennessee-Tombigbee Waterway project.

The agricultural boys are clamoring again, and their dissonance has reached the sympathetic ears of Congress and the Corps. This time, the inseparable "CC twins" are threatening the most beautiful section of Georgia's Flint River. The rea-

son: to provide flood control and to *increase production on
agricultural lands* below a $133 million dam. This latter reason,
as promoted by the Corps, might warrant a certain degree of
consideration if it were not for another very poignant fact: the
American taxpayer has already spent over $4 million annually
to keep 50,000 acres *out of production* in the same four-county
area that would be served by the dam!

This $133 million scheme, known as Sprewell Bluff Dam, is
completely lacking in economic justification. In its desperate
search to find supporting statistics, the Corps went so far as to
say "recovery and preservation of historical antiquities and arti-
facts will gain impetus from the implementation of the project." In other words, "dig like hell before we cover them with
water."

The Corps' main criterion in promoting the dam is to extol
the virtues of recreation, fish, and wildlife. This kind of reasoning becomes a bit asinine when one realizes that these are the
very things the dam would destroy. If this were not so, surely
the project would have the support, rather than the vigorous
opposition, of such state agencies as Georgia Recreation
Commission, the Georgia Natural Area Council, the Georgia
Fish and Game Commission, and the Georgia Department of
Natural Resources.

Alert conservationists may stop this project of questionable patronage. There is strong and growing support for a
scenic river and park program of the Department of Natural
Resources. If this latter plan is adopted, the citizens of Georgia
would benefit economically while preserving irreplaceable
segments of the state's environment.

≈§è≈   ≈§è≈   ≈§è≈

Every American has a vested interest in the nation's Capital; its problems are the problems of everyone. In general perspective, Washington can be considered one of the most
beautiful cities in the world, but its beauty is marred by the
black, sluggish, foul-smelling cesspool of the Potomac River.
This river—flowing through the Capital of the United States—

POTOMAC RIVER BASIN

Showing the Sixteen Major Dams
Proposed in the Potomac River Basin Report. 1963
Corps of Engineers, U.S. Army

should be the epitome, the eminent example, of sound water-management practices. Instead, it is befouled with sewage effluents and with industrial and municipal pollutants. Rather than reflecting the architecture, the landscaping, and the people of a proud city in clear, free-flowing waters, the river flows as a black-stained gutter carrying the offals of society.

Plans to correct this shameful decadence have been rampant for decades, but major and effective improvements have failed to materialize because of the inanity of most proposals. The Corps' long-range plan for the entire Potomac River Basin exemplifies this situation. Basically, this plan recommends that sixteen major dams be constructed in the basin between now and the year 2010. To understand the magnitude of this plan, the problems involved, and the alternatives available, one should have a general concept of this extensive watershed.

In addition to the District of Columbia, the states of Virginia, Maryland, West Virginia, and Pennsylvania are involved. (See location map, page 99.) Any adopted plan that is concerned with the entire river basin will affect the aesthetic, recreational, cultural, and economic values of each of these states. The basin is largely farms and timberlands, steeped in the history of early pioneers who settled in the valleys adjacent to the nation's Capital. The countryside reflects the frugality and the art of its people—contoured fields, hardwood forests, old stone and brick houses, large barns, and country churches with Christopher Wren steeples. The richest of these lands are in the low, alluvial valleys—lands that would be submerged by the Corps' big dam plan. Extensive areas of the forested Allegheny foothills would be drowned also. This history and these resources, so typically American, would be lost forever beneath the impounded waters of the Corps' implausible scheme.

To see or to smell the Potomac is assurance enough that concern for the river is justified. The basic problem is one of pollution and assuring an adequate water supply for an expanding population. A preponderance of sewage from inadequate treatment plants is the greatest cause of the pollution. In a choice bit of engineering wisdom, the Corps proposes a series of sixteen huge "toilet flushing tanks" (dams) to flush the sewage into Chesapeake Bay, rather than correct the problem at its source. The Corps maintains that the dams are essential for water supply as well as for their sewage-flushing value. Nine would be completed by 1977 and land pre-empted for the remaining seven. Also, an additional 418 "small" reservoirs would be needed for proper watershed management. The major dams, alone, would inundate a total of 78,530 acres. The acquisition of adjacent lands for "recreational" purposes would bring the total to 179,453 acres. Yet, with all these proposed impoundments, and with all the acreage involved (much of it great distances from the Capital), the main problem is still one of providing a municipal water supply and sewage abatement for the Washington area.

Considering the magnitude, the obvious fallacies, and the national significance of the Corps' plan, it would not go unchallenged. The keystone of the opposition was the National Parks Association. This is an independent, non-profit, public service organization with more than 26,000 members. This time, the Corps met with a formidable foe. Instead of just presenting vocal opposition to the plan, the association countered with an *Analysis of the Potomac River Basin Report* as presented by the Corps of Engineers, with the addition of a general alternative plan of its own. This *Analysis* had the concurrent approval of thirty-six additional environmentally concerned organizations with a total membership in excess of seven million. Much of the information presented here is based on this report.

The Corps, in announcing its initial plans to the public, reverted to its often-used tactic of disguise and delay. When the first public hearings were scheduled, they were done without detailed information being available to the public. Maps showing proposed dam sites and land acquisitions were not available until the eve of the hearings, and then only after concerted and repeated demands. In this way the Corps hoped to eliminate any intelligent and organized opposition. When the final report was made public by the Corps, it varied from the original in some instances as to the announced size of planned land acquisitions. This omission of deail was enough to have invalidated the hearings on legal grounds. Thus, by adhering to the fact that public hearings were held, the Corps precluded the hearing of new materials based on its final report.

The announced plans for extensive site pre-emption had immediate and profound effects. Business enterprises refused to expand or moved elsewhere because of the uncertain land situation. Landowners were left in a state of limbo; they did not know if, or when, or for how much, they would have to sacrifice their lands. The situation was confused further by the Corps playing a hit-and-miss game of checkers with the proposed dam sites. River Bend, the first dam proposed for the main river, was abandoned because of the overwhelming evidence of the damages that it would cause. The site was moved a few miles upstream and given a new name: Seneca. This was still to be the main dam, but its potential capacity was somewhat less than what River Bend would have been so four additional dams were deemed necessary. A reference to the map will show them as North Mountain, Licking Creek, Town Creek, and Tonoloway Creek.

In contrast to the Corps' proposed flushing techniques, the general plan endorsed by the National Parks Association and other agencies with allied interests was basically one of pollution abatement. Also, the published report provided adequate proof that a clean Potomac and its estuary could provide all the water that the Washington area would ever need. It offered new and logical approaches to flood control and recreational demands.

A positive and enlightened approach to the water problems of the nation's Capital—one based on the most modern technology—should be of interest to every American citizen. For this reason, the conclusions reached in the National Park Association's report are presented here as a summation.

SUMMARY OF CONCLUSIONS

1. The Potomac River Basin Program of the District and Division Engineers should be set aside by the Board of Engineers for Rivers and Harbors, and by the Chief of Engineers in its entirety.

2. A vigorous attack should be made immediately on water pollution throughout the entire basin. Expanded federal assistance could be given to the states and localities and to private industry.

3. Intensive pollution abatement should extend to agricultural, municipal, mining, and industrial wastes; research should be expedited in methods of detecting, locating, and removing pollution.

4. An initial plant for the distillation treatment of plant effluents with a capacity of perhaps thirty million gallons a day, should be constructed promptly at Blue Plains; pumps and a small diameter pipeline should be provided for the diversion of concentrates; an auxiliary pump and intake system for municipal water supply should be constructed below Little Falls.

5. After the initial plant, additional units of similar capacity should be constructed as the need arises for the purification of effluents; in periods of low flow the municipal water supplies being drawn from the Potomac should be supplemented temporarily by water drawn from the Estuary below Little Falls.

6. Programs for the separation of storm and sanitary sewers should be completed promptly; additional chlorination facilities should be provided at all metropolitan treatment plants, and in case of emergency at Dalecarlia.

7. Programs and facilities for the concentration of sewage plant effluents and the removal of nutrients should be developed throughout the Basin.

8. All communities throughout the Basin should be required immediately to cease any discharge of raw sewage into streams and to channel all wastes into intensive treatment plants.

9. Vigorous research should be undertaken on methods of treating municipal wastes without phosphates, or for the separation of phosphates from effluents.

10. An accelerated program of watershed management reservoirs should be initiated promptly under the Small Watershed Act, with due adherence to the requirements of local acceptance and participation by contributions of land or easements.

11. These watershed-management reservoirs should be constructed primarily for flood control and recreation, but should so be designed as to be adaptable for local agriculture, municipal, or industrial water supply purposes as needed.

12. State and federal park, forest and game lands throughout the Basin should be enlarged with the assistance of the Bureau of Outdoor Recreation and otherwise; recreation localities on such public lands should be expanded.

13. The programs of the Department of Agriculture to aid farmers in shifting crop lands to recreational use should be enlarged; the watershed protection program of the Soil Conservation Service and Forest Service should be expanded.

14. An Administrative moratorium should be imposed immediately on the programming or construction of any large dams anywhere on the Potomac and its tributaries. Such a moratorium should continue until the constructive measures outlined in this *Analysis* have had ample time to demonstrate their validity.

15. Agencies of the state and federal governments concerned with economic progress throughout the Basin should prepare programs to aid the growth of agriculture, mining, timber industries, recreational development, educational institutions, and governmental installations, to all of which the Basin is well suited, and all of which will enhance, not injure, the natural environment which is the peculiar heritage of the Basin.

16. Except for the small headwater impoundments of the watershed type, the Potomac should remain a free flowing river, living proof that a spacious and beautiful natural environment is compatible with a high material living standard of an advanced industrial civilization.

ఇ§ఏ   ఇ§ఏ   ఇ§ఏ

In a century and a half of exploiting America's waterways, the Corps has developed a passion for its sanctified position. No matter how flimsy or ridiculous the request may be, the Corps responds like new recruits at reveille. Now, in one of the most absurd and frenetic responses in its history, the Corps is heeding the call of the barkers, the hot-dog and cotton-candy peddlers, the palm readers and other fast-buck artists that purvey their wares along the boardwalk of Ocean City, Maryland. As a result, a 145 mile stretch of tidal estuaries along the coastlines of Delaware, Maryland, and Virginia is threatened.

Actually, this project was conceived in the absence of any existing problem; it is strictly a politically motivated move by the burghers of Ocean City to bring a few more dollars to their honky-tonk shoreline via increased boat traffic. Again, the project rings with the clarity of a dredge's incessant whine: the American taxpayers are being asked to sacrifice a most valuable and viable piece of their landscape at an initial cost of $13 million, and millions more for maintenance—all to pad the pocketbooks and political fortunes of a few.

The Corps' plan to satiate the demands of these politically favored few is to dig a 100-foot-wide, six-foot-deep channel inside the barrier reefs and through the tidal estuaries from Cape Henlopen in Delaware to Cape Charles, Virginia. When one considers the washing wakes of nearly constant boat traffic, and the periodic ocean storms that vent their fury along this shoreline, the problem of constant maintenance, alone, is enough to establish this proposal as being completely inane.

The Corps is completely oblivious to the far-reaching ecologic damages that such malignant ditches incur. The tidal estuaries that would be befouled by this project, combined with those of adjacent Chesapeake Bay, form the largest and most prolific fish nursery on the entire Atlantic seaboard. Damages incurred here could affect commercial and sportfishing industries from Maine to Florida. And it is impossible to dig this 145 mile ditch and remove an estimated five million cubic yards of "spoil" without imposing devastating results upon this frail ecosystem.

There are other absurdities associated with the Corps' proposal. The comparative remoteness of the area, and its protection by natural barrier reefs, make it a favorite wintering ground for hundreds of thousands of waterfowl. To despoil the food-producing chain of the area, and to make it easily accessible to hundreds of joy-riding throttle jockeys would completely destroy this refuge. Also, the federal government has already spent a minimum of $20 million to keep Assateague Island in its "natural state." This new scheme would open it to the inconsiderate hordes that have trampled other parks and historic sites into the ground. The Corps does admit, even if sheepishly so, that there would be a certain amount of pollution and ecologic damage. But it counters this with the old ruse that it is not responsible for pollution abatement and that some inconveniences are necessary to accommodate public demand. This is not a case of "public demand" but one of "the public be damned."

As a byproduct of its plan, the Corps states that the Chesapeake Bay would be available to thousands of extra boats. The reverse would be true, also: thousands of extra boats would be turned loose on the tidal estuaries. Neither place needs this additional intrusion because of space limitations and the increased pressure on ecological stability.

There is avid opposition to this grandiose scheme, and rightly so, for this time the Corps' dagger deserves to become a boomerang. There is absolutely no economic or moral justification for proceeding with this project. Even to have given it serious consideration is a shameful reflection on Congress' sense of values, for it is to blame—perhaps more so than the Corps.

…§§…   …§§…   …§§…

An incredible plan to dam a thirty-seven-mile stretch of the Delaware River simmers on the Corps' back burner. It has been simmering there since 1962 when Congress authorized the construction of Tocks Island Dam, just upstream from the historic Delaware Water Gap. The dam almost reached the "done" stage nearly a decade later, but ironically, it was the Corps' own blundering procedures that stalled the project at the last minute.

In 1955, hurricanes Carol and Edna deluged the Delaware Valley with torrential rains. This caused extensive flooding resulting in property damage and some loss of life. Although most of the damage occurred on the Delaware's tributaries, this weather phenomenon gave impetus to the idea that a huge flood-control dam was the answer to such problems. When authorization came through the passage of the Flood Control Act seven years later, the Corps was ready with its plan to inundate the valley from the Delaware Water Gap to the New York border. However, there was immediate and growing opposition to this plan that would devastate one of the most beautiful and historic valleys in the country. This opposition, and the Corps' difficulty in justifying the project economically, stalled construction plans for years. It was not until 1971 that the Corps' rejuggled figures had an irresistible appeal to the Appropriations Committee of Congress. Even though the predicted costs now exceed $300 million, more than half of this figure was absolved in abstract values for recreation and other fringe benefits.

But the Corps made one big mistake. In token compliance with the National Environmental Policy Act (NEPA), it issued an eight-page environmental impact statement for the entire project—a project that would not only affect the thirty-seven-mile stretch of the inundated valley, but one that would affect the ecology of the entire river from its source to the Delaware Bay, some 280 miles distant. Even the ecology of the bay would be affected by the irregular flows of fresh water. And the Corps tried to account for all this in a brief eight-page statement.

A small nucleus of Congressmen led by Pete du Pont of Delaware disliked the idea of the Environmental Policy Act receiving such casual treatment and made an effort to have construction funds withdrawn from the House's annual public works appropriations bill. The group was denied adequate time at committee hearings, and when the bill came to the floor of the House, its proposal for withdrawal of the funds was shouted down in the prevailing carnival atmosphere. In the July-August issue of the *Sierra Club Bulletin*, Congressman du Pont relates this action: ". . . a few young voices said 'Aye' and the 'Nos' rolled across that chamber like thunder. The lions had won, of course. And some miles upriver, the yellow bulldozers of the Corps of Engineers were fueled and ready to begin corrective surgery on the earth."

The House's blithe repudiation of NEPA and its enthusiasm for this expenditure was not shared by the Senate. Senator Clifford Case of New Jersey demanded that the Council of Environmental Quality (administrator of NEPA) have the entire project reviewed by an impartial body before any Senate action be taken. This review took an ironic twist from the original eight-page report. Now, the Corps' own consultants submitted a 100-page report that confirmed what conservationists had maintained for years: agricultural runoffs, fluctuating water levels, and other contributing factors would have disasterous effects on water quality for whatever use intended. The Council of Environmental Quality ordered that no work be started on the project until such time that all pollution factors could be controlled. Seemingly, this poses an insurmountable problem. In the meantime, the project still simmers.

꧁ఇ꧂   ꧁ఇ꧂   ꧁ఇ꧂

If there is any imbalance in the natural plan of survival for mankind, it is in the fact that supporting resources do not expand as the population increases. The quantity of water is constant; its availability, and to whom, becomes the negotiating factor. With an estimated 70% of the country's population living in the greater metropolitan areas, the demands for usable water becomes concentrated. This is most noticeable in western regions, but even in the east, the major water resources are taking on the characteristics of an African water hole. The Connecticut River is no exception.

Starting just over the Canadian border, the Connecticut flows southward for some 400 miles. It forms the boundary between Vermont and New Hampshire and continues through Massachusetts and Connecticut emptying into Long Island Sound at Old Saybrook. Nearly two million people live in the river basin, and the basin supports over 3,700 industrial plants. The current and future demands for the river's water comes not only from within the basin, but from the metropolitan areas of Boston and Providence. Even New York City is mentioned in considered proposals by certain state and federal agencies. In addition to these demands, the Connecticut is burdened with excessive pollution which must be eliminated in order to meet existing federal control standards.

In a recent survey by a number of commercial banks likely to have a financial interest in restoration plans, it was estimated that the cost to towns and industries in the four-state area would be in excess of $1.3 billion by 1980.

Citizens with a vested interest in the Connecticut Valley have long recognized the complexity of these conditions and demands. The first concerted effort toward solving the problem was made in the early 1960's by forming the Coordinating Committee for the Connecticut River Basin Investigation. This committee had a formidable array of membership from the numerous state and federal agencies that would in some way be involved or affected by any formulated plans. Probably because it had the strongest hold on federal purse strings, the Corps of Engineers became the chair agency for this body.

After six long years of study and investigation under the tutorage of the Corps, the committee's voluminous report had an old and familiar ring—dams and more dams. The proposed plan called for seven large impoundments and 118 smaller watershed projects. Once again, as on the Potomac, the Corps advanced the theory that pollution of the river could be cured by flushing it out to sea. The report contained other questionable and contradictory recommendations. It called for constructions that would interfere with existing state-owned flood-control areas. If the Quabbin Reservoir (Boston municipal water supply) were opened to additional recreation as urged, the report estimated it would cost an extra $50 million for additional treatment facilities. Yet, the Boston Metropolitan District Commission is willing to accept, and urges, that a twenty-five billion-gallon infusion of Connecticut waters from other recreational areas be diverted into the reservoir each spring. If the water is good enough for the residents of the Connecticut Basin, it should be equally palatable for those in Boston. Is the $50 million figure another diversionary tactic to ignore public requests?

One of the most visionary and commendable aspects of the report is perhaps the most contradictory. In dreams of yesteryears, the plan calls for establishing a sustained run of 40,000 Atlantic salmon by stocking one million smolts annually. But in complete contradiction, this same plan would inundate a large portion of the only suitable spawning habitats. Similar conflicting presentations were made relative to the development of electrical power and the volume of water needed to do so.

Since the Coordinating Committee was dominated by the Corps, the need to challenge the inefficiencies of the plan would have to be championed by an impartial group. The Connecticut River Watershed Council assumed this role and continues to lead the fight for a more equitable and efficient solution to the problems of the river basin.

This organization submitted a board of director's review of the overall plan to the Coordinating Committee on February 11, 1970. This review made the following points regarding the controversial issue of power:

> Reference has been made in the report as to the ability of doubling the Haddam Neck nuclear output and tripling that of the Vernon plant. Our communications with the power officials have indicated that there are no plans or intentions of expanding the power outputs of these plants. The parameters of the flow of the river, its volume, and other environmental considerations militates against increasing of these plants at the present stage of technology, according to these officials.
>
> Studies have been made which reveal that cooling towers, used by the nuclear plants for reducing the temperature of waters, may reduce the volume of water by as much as 20% depending on climate and geographical location. Most nuclear plants in the valley require 6000 gallons of water a second for cooling. In the Vernon plant, where cooling towers will be used, the water loss through evaporation may average 3% (according to company information) which would mean a water loss of 648,000 gallons every hour, or 13 mgd or more depending upon external temperatures. This is a significant amount of water—especially in this stretch of the river, and it will occur primarily during the summer months when river flows are low. This figure may be considerably higher after the plant goes into operation, but even at this rate the river could not tolerate a tripling of the electric output at Vernon as noted as feasible in the summary report.

Such critical facts as these were ignored by the Corps in its determination to adhere to the original report. The Corps was determined to supply water to metropolitan Boston, even though its report failed to show that surplus water was available in the watershed. Obviously, this inclusion would satiate more wide-spread political interests—and the proposed pipeline would carry pork-barrel money as well as water.

In its attempt to comply with the legally required impact statements, the Corps tried to bluff its way with generalizations and by ignoring the findings and requests of public interests. This was especially true in the area that would divert water to Boston. The Connecticut River Watershed Council continued to challenge these statements on behalf of the citizens in the river basin. On December 13, 1972, John C. Calhoun, Jr., the Council's president, wrote the following letter to the Division Chief of the Corps of Engineers. It once again reveals the citizens' concern and response to the Corps' impact statements. It again asked some pointed questions of specific concern to everyone in the valley.

December 13, 1972

John Wm. Leslie, Chief
Engineering Division
Department of the Army
New England Division, Corps of Engineers
424 Trapelo Road
Waltham, Massachusetts 02154

Re: Preliminary Draft Environmental Statement
     Northfield Mountain and Miller's River Basin Water
     Supply Projects

Dear Mr. Leslie:
     The Connecticut River Watershed Council, Inc. has participated in early discussions and formal public hearings relative to both the above out-of-basin transfers of water from the Connecticut River valley. We have submitted formal statements on these proposals to appropriate state and federal agencies and the Boston Metropolitan District Commission, and including the General Court of the Commonwealth beginning in 1969. We would call your attention specifically to our statements filed with Colonel Frank P. Bane dated February 11, 1970 and July 6, 1972 which are concerned, either in part or

in whole, with the diversions of water either from Northfield Mt. or the Miller's River.

We have repeatedly raised questions which represent the concerns of our valley's residents during public hearings on these projects. Moreover, we have requested the U. S. Army Corps of Engineers, or the Boston Metropolitan District Commission, to provide answers to and action on those matters of concern expressed in our submitted statements. The fact that no attention to the views and concerns of the valley citizens is evident either at hearings, follow-up correspondence, or in succeeding published reports is indicative of the lack of concern your agency has for public response.

The current draft environmental Statement is further testimony of the U. S. Army Corps' disinterest in responding to private citizens' and organizations' views which have been expressed repeatedly at earlier meetings and public hearings. The Statement also substantiates the tenor of the federal General Accounting Office's recent report which criticizes the Corps of Engineers, among others, for not disclosing fully all adverse environmental effects of their projects; by complying with the 1969 law on national environmental policy in form but not in substance, and that, in general, the quality of the impact statements is poor.

In this respect, we wish to submit the following comments and to raise again several specific concerns as our response to the Draft Environmental Statement.

(1) We believe the entire diversion projects have been based upon an unacceptable premise—which is they are predicated upon a control flow of a minimum of 17,000 cfs in the Connecticut River at Montague City U. S. Geological Survey gaging station. This control flow was elected not from any environmental determination but rather from Massachusetts State Legislation (relative to the Northfield Mountain Project) which was based solely upon the hydro-power generation needs at Turner's Falls dam. The 17,000 cfs flow, which includes the peak average annual flow of the Deerfield River, is principally the storage head needed at Turner's Falls to develop hydro-power at the Cabot station.

To use the needed make-up water for power generation as the desirable minimum flow for diversion purposes, is certainly not responsible environmental determination. This is particularly undesirable in consideration of the Miller's River diversion which is also set for the same stage flows at the Montague gaging station.

As increased water flows are being considered by the

Federal Power Commission in relicensing of Turner's Falls and the four immediate upstream dams (currently an additional .20 cfs per square mile of drainage), it would be consistent in the Corps of Engineers' thinking to make these additional flows available for diversion under the current method of streamflow calculation—although, in fact, these increased flows are for the enhancement of anadromous and resident fish species and recreation in the Connecticut River.

Therefore, we believe that another premise must be found upon which to determine possible environmental impact of water diversions rather than that based upon hydroelectric power needs.

(2) Our Council has questioned many times the statements made by responsible state and federal agencies that there exists a 'substantial surplus of water' in the Connecticut valley that will not be needed for valley use. Never has this been substantiated in reports issued by the Corps of Engineers.

We do not believe that the availability of water (surface and ground supplies) and the future demands for water supply in the Connecticut River valley have been fully studied or planned for. Until these are ascertained, no judgment of value can be made as to the effects of water diversions on the donor area.

We suspect that the lack of consideration given in this Statement to the environmental impact of diversions on the donor area is a direct result of not having this basic and critical information available.

(3) It can also be *assumed* that identical impacts may occur within the donor area if diversions are allowed as has been suggested to occur in the receiver area if diversions are not allowed. It is likely that as additional water supplies are needed in the valley—both in Massachusetts and Connecticut—that a 'prior appropriation' of these waters outside the valley will result in a slow-down in this valley's socio-economic development, municipal services, employment, recreation opportunities, and frustration of citizens as "they see their lawns dry up during watering bans."

(4) Since 1970, the Council has consistently asked the following questions with no answers having been given. We ask these again:

a). What legal commitments are to be provided by the MDC for returning proposed diversion waters when they are needed by riparian communities, landowners, and industries within the valley in Massachusetts and Connecticut?

b). What compensation, or equities, will be established

with riparian communities, landowners, and industries for the withdrawal of their riparian waters? We see no "quid pro quo."

c). What assurances are being provided that present proposed diversion levels will not be exceeded after 1990, or before?

We believe that these questions should be answered now by the Boston Metropolitan District Commission and made part of the final environmental statement report.

(5) The report dismisses the development of the substantial (300 mgd) groundwater resources in Plymouth County as being more costly than Northfield and the Miller's projects combined—yet, we have no costs to compare for any of the projects.

Although the report gives scant reasons for dismissing further consideration of groundwater as an alternative, we disagree with those given, and urge that more in-depth, impartial, evaluation be given to this alternative.

(6) Except for a 2½ page description of facilities, there is no impact evaluation of the actual construction of the various aqueduct tunnels. CRWC has already experienced that such activity is detrimental, based on M.D.C. field investigation activities in the Wendell section of the Northfield-Quabbin line and in particular at our Whetstone Wood Sanctuary in that town.

Privately, M.D.C. officials have suggested that there is a possibility of "temporary" disruption of private wells (groundwater) as a result of the tunnel itself. Further, the so-called construction shafts will require a surrounding land area for support facilities and to provide a place to stockpile waste rock from the tunnel. Such sites will doubtless be areas of intense heavy equipment activity. Trucks and machinery will be active in and about these sites, and the impact will be felt in the forms of additional noise and air pollution in the area, as well as adverse impact on existing light duty road facilities which could be degraded by heavy trucking.

The volume of rock resulting from a tunnel 9.8 miles long, 10 feet in diameter, is considerable, yet no mention is made as to where this material is to be placed. The final permanent location of the spoil rock should certainly be discussed in the impact statement.

(7) Before large water diversion projects with substantial long term capital investments are constructed in this valley, we submit that much more study and evaluation of alternative water supply sources in eastern Massachusetts needs to be

done. And, we believe that further environmental impact studies are needed on the donor area in the instance of the Northfield-Miller's River proposals.

We believe this response to the November 1972 Draft Environmental Statement is clear, and that the time is at hand to have questions answered. We respect your decision that our response will be included as part of the Final Statement.

> Respectfully submitted,
> John C. Calhoun, Jr.
> President

To this, the Corps' response has once again been to hell with you, Jack, we'll do things our way.

CHAPTER 6

Midwest Madness

AS THE CORPS' menacing tentacles touched down across the great expanse of central farmlands, every state and every major waterway cringed before the monster that would lay devastation upon the land. In the eyes of the Corps, the natural environment existed as a panorama of ill-arranged stupidity. Its obsession for "molding the earth to meet the needs of man" subjected the Midwest to this unrestrainable madness. No area was too scenic, too historic, or too valuable to escape the Corps' dam-building craze.

All the Corps' projects seem to be blessed with the same degree of absurdity, so it becomes somewhat difficult to say that one is more inane than another. But surely, the proposed Oakley Dam on the Sangamon River would vie for top honors.

The Sangamon winds its way across central Illinois, past Decatur and Springfield, enroute to its junction with the Illinois River. The Oakley Dam is the brainchild of politicians and the Corps. It is now referred to by the Corps as "Springer Dam," a tribute to the former Congressman, William L. Springer, who continues to fight for the project. The completion of the dam would destroy Allerton Park, a 1,500-acre tract of mostly virgin woodlands held in public trust by the University of Illinois.

Allerton Park was given to the University in 1946 to be used "as an educational and research center, as a forest and wildlife and plant life reserve, as an example of landscape architecture, and as a public park." The park woodlands have grown undisturbed for some 20,000 years and is the University's most valuable outdoor laboratory and ecological research center. In addition, the park has been made a National Natural Landmark, but such designations do not destroy the designs of the Corps.

The Sangamon Valley echoed with the strains of the Corps' same old theme: flood control, water supply, and recreation. The approach to meeting these objectives was also the same: dams dilute pollution by flushing and replace all forms of recreation with that of boating. The Corps' adherence to familiar and time-worn procedures again revealed its hit-and-miss engineering genius. In 1961, the Corps presented a plan for a forty-nine-foot dam twelve miles downriver from the park. The 645-foot (above sea level) flood level would inundate a large portion of the park. Also, an additional 2,800 acres of downstream land would be subject to clearing and 100 miles of channelization. This plan was authorized by Congress in 1962, but by 1966, the Corps recognized its own miscalculations. It had underestimated potential siltation and had failed to base its findings on maximum flood records. The answer: a bigger dam—one sixty feet in height with a flood level of 654 feet. Close to half the park would be under water during flood periods. Three years later the Corps maintained that the only way it could meet the state's new water quality standards was by increasing the size of the proposed dam. Now, the flood level would be 656 feet, exceeding Congressional authorization, and leaving only the high ridges of the park exposed.

Of course, such bungling of plans for this unique ecological area did not escape the attention of conservationists and others with an interest in the park. The public's response was the formation of the Committee on Allerton Park. The Corps soon learned this was not a group that could be blithely labeled as "butterfly chasers" and then ignored. This was a group of professionals—scientists, engineers, economists, and lawyers. In opposing the project, this committee did not follow the usual format of objecting on just aesthetic and conservation grounds. Instead, the committee challenged the Corps at its own game. Engineers out-engineered the Corps by revealing the omissions and the fallacies of its plan, and by presenting alternatives that would be more economical, save Allerton Park, and impose a minimum of disturbance on the environment as a whole. Economists debunked the padded benefit-cost ratio; lawyers challenged the legal aspects of the whole project; and committee workers presented the Illinois Congressional delegation with a petition of 80,000 signatures in opposition to the project.

The Committee on Allerton Park published its findings in a ninety-three-page booklet entitled *Battle for the Sangamon.* Excerpts from the booklet's summary follow (with the committee's permission), because they present a vivid professional analysis of the Corps' adroitness at foisting its ill-conceived schemes upon the public:

Water Supply

1. An abundant supply of pure water can be obtained by Decatur from the Mahomet Valley Aquifer at far less cost than that from the reservoir. The well field could be developed in stages as water needs develop. The present dam proposal contains more than twice as much water supply storage as the original proposal, even though no city has agreed to buy the extra water.
2. The nitrate content of Lake Decatur's water is increasing and has exceeded the United States Public Health Service limit of 45 mg/1. The Oakley Dam is to be built in Lake Decatur. The reservoir will not be able to continuously supply water that meets USPHS standards.
3. Lake Decatur has lost over 35% of its water storage

capacity because of sedimentation since it was completed in 1922. The Oakley Reservoir will serve as a silt trap and is expected to collect 100 acrefeet of sediment per year. This reservoir is being designed to store 10,000 acrefeet of sediment and only 7,000 acrefeet of water.

4. The Sangamon River carries excessive quantities of nutrients such as nitrates and phosphates. These chemicals will cause algae to grow rapidly and foul the water in the Oakley Reservoir. Lake Decatur presently has an algae problem.

Low Flow Augmentation

1. The present proposal includes 9,000 acrefeet of storage to maintain a two-feet water depth in the Sangamon River below Lake Decatur. This storage is unnecessary and will cause drainage problems in farm fields around the Friends Creek Reservoir.

2. Although Decatur has publicly agreed not to use Oakley's water to dilute sewage effluent, the specter of such use remains. Decatur's new sewage plant is being built to handle the effluent load expected by 1972. After that time the Illinois Division of Waterways plans to dilute effluent with water from the dams.

3. A direct result of low flow augmentation will be mudflats. Mudflats form when water is released exposing the reservoir bottom. Mudflats hamper recreation and are malodorous.

Recreation

1. With the exception of motor boating and water skiing, all forms of recreation claimed for the Oakley Project can be realized without building dams.

2. Within sixty-five miles of the Oakley site there are 26,000 acres of public water surface and only 3,500 acres of public woodland. Over one-third of this woodland is in Allerton Park.

3. The high degree of pollution present in Lake Decatur is expected to exist at Oakley. Mudflats and pollutants will make the Oakley Reservoir unfit for swimming and many other activities.

Flood Control

1. The reservoirs will end production of three times as much land as they will protect in the average year.

2. The project is expected to reduce annual urban damages by only $14,000. Farmland which the Corps proposes to protect can be protected at far less cost by using existing federal crop insurance and/or levees. Bottomland prices are lower because of periodic flooding so that farmers there realize the same rate of return on an investment as do upland farmers.

3. The Carlyle Reservoir in southern Illinois has actually increased flood damages above and below the dam. The same could happen at the Oakley Reservoir.

4. A greenbelt of 21,000 acres along the Sangamon River is to be purchased by the Corps to carry flood water releases from the project and provide recreation. The greenbelt is less costly than the originally planned "channel improvement" and will do far less ecological damage.

Economics

1. The Corps claims that the benefit-cost (B-C) ratio for the project is 1.15. This means that in the average year $1.15 in benefits will be returned for each dollar in costs.

2. The Corps made a mistake in its B-C analysis when it figured the cost of an alternative water supply for Decatur at an interest rate different than that used for the project. Correcting this mistake brings the B-C ratio down to 1.05.

3. The Corps made another mistake when it included the greenbelt in land upon which it claimed flood reduction benefits. Correction of this error brings the B-C ratio down to 0.97, meaning that benefits are less than costs.

4. The Corps has based its B-C analysis on an interest rate of 3.25%. The current interest rate for Corps projects is 5.125%. Recalculating the B-C ratio at 5.125% and correcting the two mistakes brings the B-C ratio down to 0.69.

5. The Corps has claimed benefits for swimming and water quality that will not be realized. The Corps should also have used well water in its B-C analysis as the least cost alternative water supply. Correcting these water matters brings the B-C ratio down to 0.49.

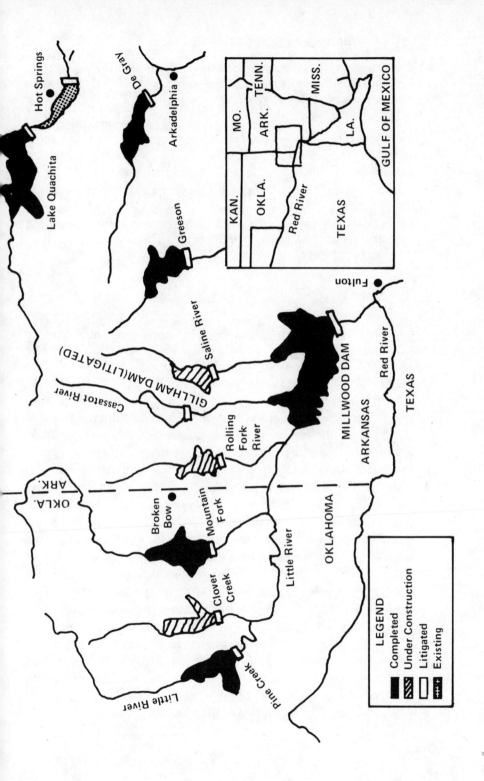

6. Considering all the pressing needs facing federal and local governments and the general lack of money in government treasuries, proceeding with the low priority Oakley Project at this time makes no sense. The project presents planners with an easy way to save $65 million.

Alternatives

1. Almost all of the benefits claimed for the Oakley Project can be realized by treating the Upper Sangamon watershed for erosion control, developing the greenbelt, and using well water for Decatur. The cost of this alternative project would be about $26 million.

Legal

1. The present Oakley Project violates several federal laws including the National Environmental Policy Act of 1969, the Flood Control Act of 1936, and the 1962 law authorizing the original project.

In the years of maneuvering to foist its unwanted plan upon the citizens of Illinois, the Corps continued to indict itself before the expertise of the Committee on Allerton Park. Perhaps its most flagrant legal violation was the attempt to prevent public hearings by withholding regulations on procedures. The threat of legal action made the hearings possible, and within a short time numerous alternatives to the original plan were being evaluated publicly.

Recent reports from the committee indicate the Corps has been so soundly defeated that the project has been turned over to the state. An alternative plan proposed by the state's engineers would protect Allerton Park, keep the Sangamon River from being used for sewage dilution, and provide an expansive recreational greenbelt instead of destructive channelization. This plan, although a compromise, has won the support of most groups, including the Committee on Allerton Park and the University of Illinois Board of Trustees. However, the committee has warned that any attempt to enlarge upon the plan will bring immediate and increased opposition.

The state of Arkansas has been subjected repeatedly to the dam-building frenetics of the Corps, especially the southwestern corner where numerous river basins empty their waters into the Red River. A number of dams have been completed, some are being constructed, and others are still in the planning stage. Two projects, the Millwood and Gillham Dams, (see location map, page 121) again illustrate the Corps' chicanery in promoting its dubious plans.

In 1946, the Corps recommended to Congress the construction of Millwood Dam on the Little River to create a reservoir with a capacity of 2,208,000 acrefeet. To do so, some 107,000 acres of valuable hardwood forests would have to be inundated. The Corps let the matter rest at this point and did not request funding of the project; this proved to be the initial step in a long and devious plan.

In 1950, the Corps went to Congress with the story that in order to save some hardwood timberlands for a commercial lumber mill at Broken Bow, Oklahoma, the selected location for the Millwood Dam should be moved. When the Corps' full machinations were finally revealed, it turned out that instead of a slight alteration for the site of Millwood Dam, the Corps had come up with an entirely different reservoir system. It had parlayed the one original Millwood Dam into seven dams! Also, instead of an alternate site for Millwood, the Corps now proposed that it be built at the original site, but with a capacity reduced by 25%. Thus, with eventual Congressional approval, one authorization that lay dormant for years was now expanded into a seven-dam proposal.

In an attempt to find support for this all-inclusive plan, the Corps devised a benefit-cost ratio of 1.47 It based this upon getting a dependable daily yield of 621 million gallons of water. It is difficult to understand how even the far-reaching influence of the Corps could have sold this grandiose scheme to Congress. Such a quantity of water was forty times greater than the 15.3 million gallons daily (mgd) actually *used* in the region. The 621 mgd was far greater than the region had the need, desire, or the ability to pay for. Yet, the Corps contended that the unattainable yield from this water was justification for a 1.47 benefit-cost ratio—a completely erroneous deduction.

The Department of the Interior had this to say about the project, as recorded on page 12 of House Document 170:

> Reviewed in the light of requirements of the Bureau of Reclamation in the Department of the Interior, the report does not clearly show the need for water, a plan for use of the water, economic justification and feasibility, repayment willingness, and contracting ability of potential water users, or project water rights.

Incredibly, the water generation proposed by the Corps with this system of environmentally devastating dams is by official figures eight-times the 1980 requirements of the region, and an unbelievable four-times the need of 2080—more than 100 years from now!

Nevertheless, the Millwood Dam was completed, and the inundated area was 100,000 acres—a reduction of only 6% not 25%. Quite obviously, the hardwood lumber mill at Broken Bow had served its purpose and now had been forgotten.

One of the six additional dams proposed by the Corps was to be on the Cossatot River. For some questionable reason, this dam was not identified by name in the Corps' presentation to Congress. Eventually, it became known as the Gillham Dam. The Corps attempted to justify this dam as a flood-control measure for the Red River. Actually, there was no need to impound the waters of the Cossatot because the already completed Millwood Dam, plus the backup dams of Pine Creek and Broken Bow (see map on p. 121), were adequate to control flood waters.

But the Corps blundered onward, once again giving only token efforts to the required environmental impact statements. This was a big mistake. Wellborn Jack, Jr., an astute attorney from Shreveport, Louisiana with a personal love for the Ozarks, brought suit against the Corps on behalf of the Ozark Society and the Environmental Defense Fund. In the United States Eastern District Court of Arkansas, he charged the Corps with ignoring agencies with a vested interest in the

Cossatot and with not filing proper impact reports. The court ruled for the plaintiff and stopped all work on the Gillham project until proper impact statements were made available to interested agencies.

Again, the Corps tried to bluff its way; the new impact statement was basically a rehash of the old one, a fact that was quickly recognized by United States District Justice Garnett Thomas Eisele. The few excerpts from Justice Eisele's ruling which follow illustrate the Corps' blatant lack of concern for the public's interest in this and similar projects. Justice Eisele found:

> The statement did not set forth all the environmental impacts which are known to the defendants by their own investigations, or which have been brought to their attention by others.
>
> The statements do not adequately set forth the "adverse environmental effects which cannot be avoided should the dam be built as planned."
>
> The statements do not adequately bring to the reader's attention all "irrevocable and irretrievable commitments of resources which would be involved in the proposed action should it be implemented."
>
> The evidence does not indicate that, prior to making the statements, the defendants did "consult with and obtain the comments of" all federal agencies which have jurisdiction by law or special expertise with respect to any environmental impact involved.
>
> The court in its Memorandum Opinion Number Four discussed the deficiencies of the first impact statement. On January 22, 1971, the defendants filed a new or "updated" environmental impact statement with the Council of Environmental Quality. This statement is in evidence as Defendant's Exhibit. *It is not substantially different from the original impact statement* which was filed on October 5, 1970. In making of the new and original impact statements, *the defendants simply "recast" the information already in their files.* The testimony of the defendants is very candid and forthright in this respect. They felt that such an approach was the only "practical" means of handling impact statements for this type of project. No attempt was made to consider environmental consequences

that were not fully considered or evaluated previously during the life of the project. Furthermore, in revising their environmental impact statement, the defendants did not even take into account the contentions of the plantiff as reflected in the testimony in this court as the preliminary hearing on Nov. 24 and 25, 1971. *No understanding of the issue raised in this suit can be gleaned from a reading of the new statement.* (Italics are the author's.)

(The Corps frequently has voiced the claim that it builds only what the Congress has authorized and funded, and therefore, what it does is just. Justice Eisele included a communication from Congress which does much to nullify this long-used excuse.)

The report of the Senate Committee on Appropriations on H. R. 18127, 91st Congress, 2nd Session (S.Rep.-No. 91-118, 91st Congress, 2nd Sess.) covering *inter alia*, the defendants "civil" appropriations for the year ending June 30, 1971, states in part: "The committee has received objections based on environmental grounds to many programs and projects for which funds are included in this bill. The objections are principally based on the failure of the agencies involved to file the five-point statement required by the National Environmental Policy Act of 1969. The agencies were given until June 1, 1970, to prepare their procedures for implementing that act. The committee has been informed that the required statements are in preparation. In most cases the projects objected to have been under construction for some time. The fact that the committee has recommended funds in this bill does not exempt the construction agencies from complying with the provisions of that act as currently applicable."

With the bulldozers and the draglines quiet along the Cossatot, the Corps is working on still another impact statement—one it hopes will justify the building of another unwanted dam. Decisions such as that made by Justice Eisele should prove one thing to the Corps: the National Environmental Policy Act does have teeth in it—and the public intends to enforce it.

The words "environmental impact" have become a nemesis to the Corps. Numerous projects have been stopped because of its lack of compliance. One, the plan to channelize the Cache River and the Bayou DeView in Eastern Arkansas, was halted by court order in March 1973. Three months earlier, the 8th Circuit Court of Appeals in St. Louis ruled that the Corps failed to meet the legal impact requirements, but supporters of the project appealed to the Supreme Court.

The environment of the Cache River region has a very special significance: it is the largest wintering area for mallard ducks in North America. Annual flocks total from 500,000 to 800,000 birds. In addition, other species of ducks, wild turkey, deer, and small game are at home in this last significant stand of delta hardwood forests in eastern Arkansas. The Corps, entranced with its "channelization syndrome," is oblivious to such features. It plans to drain more than 170,000 acres with 232 miles of ditches. This flagrant intrusion upon a singularly unique piece of the American landscape is supported by selfish interests and their political allies who hope to profit from the decimated forests and "new" agricultural lands. Again, the unsuspecting taxpayer is being asked to pay the bill of $73 million, while they dole out millions more to keep farmlands idle! Surely, there is something wrong with the Congressional thinking that expects the American people to support this kind of stupidity.

◆§◈▸   ◆§◈▸   ◆§◈▸

There is a hint of irony in the fact that the Corps' selection of a dam site is invariably the most scenic and naturally historic section of any river. In reality, it is the gorges, the ravines, and the valleys that are most adaptable to engineering designs. For this reason, the Corps' impoundments usually destroy more aesthetic, historic, and recreational values than they create. Such is the case with the Meramec River in Missouri.

The Meramec winds its way from the Ozarks to St. Louis where it joins the Mississippi. If the Corps has its way, Meramec Park Dam will be built sixty-five miles upstream from St.

Louis. This $87.5 million dam would flood forty-three miles of the most scenic portion of the river.

The Corps' pursuit of this project has a long and tenacious history. Its original authorization dates back to 1938, but the Corps has an uncanny knack of setting on a single prerogative until it hatches into a whole brood. A case in point occurred in 1966 when a parental Congress authorized a total of thirty-one dams for the Meramec and its Ozark tributaries.

Attempts to justify the Meramec Park Dam border on the comic and the ridiculous. Originally, the reasons touted were for flood control and water augmentation for navigation on the Mississippi. These reasons are no longer considered paramount; recreation is now the chief criterion. This kind of thinking is indeed pathetic. Obsessed with the idea that boating and recreation are synonomous, the Corps employs the term liberally to attract sympathetic support for its projects. Recreation (boating) cannot be considered a justifiable excuse for building the Meramec Park Dam for the following reasons:

- The Meramec is already one of the most popular "float" rivers in the Ozark region. Hundreds of families use it annually.
- The state of Missouri already abounds with lakes and reservoirs offering boating opportunities.
- An estimated 80% of the fish species native to cool, free-running waters would not survive in the warm, impounded waters.
- Two of the finest smallmouth bass rivers in the country, the Huzzah and the Courtois, would be partially inundated.
- Three species of freshwater clams would be endangered by the project.
- About 100 caves and 350 archeological sites would be flooded.
- There would be a tremendous loss of wildlife habitat.
- High barren shorelines and mudflats will be exposed during much of the year making recreation undesirable.
- The Corps' plan does not provide for any urban recreation, the place where facilities are needed most.

• Hiking, exploration, floating, canoeing, stream fishing, hunting, and similar sports now provide the recreational needs for more people than would be attracted to flat-water boating.

The Corps' method of figuring recreational values can be looked upon as an obvious attempt at deceit. Otherwise, how can the Corps list the recreational aspects of the Meramec project as 28% of the total benefits, and refuse to put a figure on the loss of all the activities listed above?

No matter what benefit-cost ratio the Corps finally comes up with, it will certainly not warrant a $87.5 million differential. Actually, the cost of this project cannot be determined with any degree of accuracy. The area is one of karst limestone, full of cracks, ravines, caves, and underground springs. The Corps admits that extensive grouting will be necessary but does not know the cost. It could be astronomical.

But the Corps is in trouble; its old nemesis is bugging it once again. An eight-page impact statement was not acceptable to environmentalists, so the Corps is in court trying to defend it. Hopefully, the court, too, will find it unacceptable.

꿍ᢒ᠍᠍᠍᠍᠍   꿍ᢒ᠍᠍   꿍ᢒ᠍᠍

Texas is big, and Texans have a reputation for doing things in a big way, but the Corps of Engineers is determined to show the people of this great state just what the word "big" really means. Its current waterway plans make all previous projects look like so many street puddles after a rain shower. But with the Corps, size and absurdity grow together.

The Corps has had a long tenure in Texas, especially in the Galveston Bay area. This estuary is the largest and the most ecologically significant of those along the entire Texas coastline. More than 500 square miles of alluvial flatlands support hundreds of thousands of waterfowl, shorebirds, and other species associated with a brackish environment. Thousands of ducks and geese from Canada and the northern states winter in the estuary annually. It is the nursery for many species of fish; it supports a $200 million yearly sport and commercial fishing industry. Thus, the ecology of the bay area affects food chains from the Gulf of Mexico to the Arctic. But this great estuary is changing; pollution, and dredging and filling for navigational and land development interests, continue to destroy it for the benefit of a few. Current projects threaten its survival as a viable estuary.

Problems for Galveston Bay began more than fifty years ago, and the Corps was deeply involved. It straightened the Buffalo Bayou, a stream flowing from Houston into the bay, into a fifty-mile-long ditch known as the Houston Ship Channel. This project eventually made Houston one of the largest shipping ports in the country, but the channel became an industrial sewer emptying into Galveston Bay.

Now the Corps has bigger and fonder dreams of doing the same thing for the Fort Worth-Dallas area. The plan is the biggest and most expensive of all the Corps' projects. At a cost of $1.6 billion, it plans to turn the Trinity River into a 335-mile-long ditch so the Fort Worth-Dallas area can become a seaport for the barge industry. When the Corps talked "big" it was not joking; plans for the Trinity River Canal call for sixteen navigation dams, five reservoirs, and twenty locks, all of which would alter the water capacities of thousands of square miles.

Once again the Corps is trying to deceive the public by touting the cheapness of water transportation. But cheap to whom? Ironically, the taxpayer is being asked to foot this tremendous bill—plus the cost of *operation and maintenance!* Surely, every industry in the country would love to have this kind of financial backing, but this is obviously preposterous. So why should one particular industry be favored with such generous public patronage? If the barge industry had to pay for the construction, operation, and maintenance of this project, water transportation would not be so cheap, and economic justification by the Corps would be a joke. Rail and highway transportation from the Fort Worth-Dallas area could be built and maintained at a fraction of the cost for the canal.

The effects of another industrial sewer emptying into Galveston Bay is difficult to evaluate. The ecology of the bay is already at the critical point, and the effluents from this canal, plus the barrier impoundments, may be sufficient to completely ruin the estuary. The Wallisville Barrier Dam, the first step in this massive project, has been started across the mouth of the Trinity River. Completion of this dam (it is being challenged in court) would eliminate 12,000 acres of the estuary.

There is a definite conflict of interest in the whole project. Local voters in the seventeen-county area that would be affected expressed their disapproval by defeating a bond issue to finance local costs. But support continues from those who would profit. A statement released in *Disasters in Water Development* (a report sponsored by American Rivers Conservation Council and other environmental groups) makes this understandable:

> One third of the Board of Directors of the Trinity River Authority, which is the state agency promoting the project, are landowners along the river. Another big landowner along the river is the Southland Life Insurance Company and some of its Board of Directors are also on the Board of the Trinity River Authority. No wonder they are eager to get this $1.6 billion project.

Work on the Wallisville Barrier was halted by court order in February 1973 for the lack of a proper impact statement. The court's decision implied the Corps had been ". . . less than objective by engaging in rationalizations and super salesmanship."

In addition to pollution and the monumental Trinity River plan, Galveston Bay is threatened with other schemes devised and promoted by the Corps. The most injurious of these is a $600 million plan to dike the perimeter lands of the bay to prevent hurricane flooding. This would pave the way for filling, land speculation, and industrialization. But more significantly, it would cut off much of the normal salt water flow from the Gulf, thus changing the ecology of the bay.

Dredging, filling, and channelization are daily Corps activities in the Galveston-Houston area. More than $100 million of taxpayers' money have been spent on "little" projects, and another $56 million in backlog appropriations are available for use, thanks to a politically sympathetic Congress.

The most spectacular of all the plots ever devised to

manage Texas waters is the Texas Water Plan—the colossal dream of the Texas Water Development Board and the United States Bureau of Reclamation. This politically controversial scheme, in conjunction with current and proposed Corps projects, could completely change the biological climate of the entire Texas coastline. Basically, it is a plan to transfer waters from the wet eastern sections of Texas to the semi-arid lands of the plains and the lower Gulf coast. It would, in effect, dam virtually all of Texas' free-flowing streams into one giant coastal aqueduct.

The Texas Water Plan is so comprehensive, both physically and politically, that it is not easily understood. It is being released to the public gradually in diluted form. When the political climate is right, Americans all across the land will be asked to pay for this multi-billion dollar blunder.

There is an element of sadness in the gray-greens, the tans, and the browns of a spreading timberless estuary. To the industrialist, the shopkeeper, the land speculator, and the man in the street, it is not a place of enthralling beauty. To almost everyone, an estuary is lacking in aesthetic appeal.

Yet, there are those who do know and understand the importance of such areas. The ecological loss of Galveston Bay, and estuary after estuary all along the Gulf and ocean coastlines, will have profound and adverse effects upon the quality of life for all people.

CHAPTER 7

Far West Fiascos

THE HORIZONS OF the Far West have beckoned the Corps ever since its first scratchings in the Sacramento Basin nearly a century ago. The arid desert lands, the high mountain valleys, the scenic gorges, the wild tumbling rivers—these were irresistible challenges to the engineering genius of the Corps. Did these wonders not exist only as nature had made them? And was it not the innate responsibility of the Corps "to mold the land to suit the needs of man?" And so it was that year after year, the Corps spread its plague of destruction across the vast vistas of the West.

The plague was contagious and malignant and spread through the veins of the land as river after river was stilled and clotted with the dams of the Corps. Here, in the West, the rivers are big and wild. To the Corps, each rapids was a waste of energy, and each gorge a natural reservoir that needed a retaining dam. No project was too big, nor any excuse too flimsy.

And conquer the West, the Corps did. But in doing so, it left a new trail through this western wonderland—a trail of horrors that not only altered the physical aspects of the land, but a trail that imposed an unwanted way of life upon many of its people. The damming of the West is an unsavory monument to a complacent Congress.

In the West, as elsewhere across the country, some of its people must share a portion of the blame for this random assault upon water resources. They have not only supported a permissive Congress, but they have zealously promoted schemes of expansion in defiance of physical and geographic limitations.

The building of cities in areas of limited water resources is an example of defying limitations imposed by geography. Invariably, in cases of this sort, there are other physical and social features that attract people; a population boom follows and the struggle to find enough life-sustaining water begins. Southern California is a case in point. Most of the water for Los Angeles, San Diego, and other cities in the southern part of California comes from hundreds of miles inland. Other states have challenged California's right to depend on them for its source of water. This battle for water rights will mount as the population increases and the availability of water lessens.

Arizona is a good example of how promoters are creating their own demise. The state's enchanting beauty, its warm healthful climate, and the mineral-rich farmlands are being used as propaganda to lure more people and industry to this land of desert and sunshine. Propagandists fail to mention the fact that Arizona is approaching a catastrophic future due to insufficient water. Nearly three-quarters of all the water used in the state comes from wells. Water tables are dropping rapidly as this underground supply is being used up twice as fast as it is being replenished. This practice has a positive and predictable end.

Without a doubt, Arizona must either find additional sources for its needed water or impose a harsh priority system for use of the current supply. How long can it afford to spread nine-tenths of its total water intake across the hot arid lands in order to bring early-crop fruits and vegetables to northern markets?

For too long, Arizona has turned toward the Colorado River, but so have the states from the Rocky Mountains southward into Mexico. If all the schemes to utilize the water of the Colorado were suddenly put into effect, the demands would far exceed the river's total volume. Now, Arizona must look elsewhere to quench its growing thirst. Cautiously, it extends an empty cup toward the Pacific Northwest. If it is ever to be filled with water, it will have to overcome the political rattlings of a dozen other cups reaching in the same direction.

Natural water distribution in Arizona is singularly unique; water quantities are limited, and their dispersal is controlled by the harsh characteristics of the land. It is safe to estimate that from one-third to one-half of the state is desert or semi-desert in nature. Because the desert areas have many mountain ranges running through them, small rivers form a lacework throughout and make irrigation possible. The deserts receive little rain; most of it falls on the higher mountain regions. When the rains are heavy enough, water will flow in the desert streams such as the Simon, Pedro, Santa Cruz, and others as it did before heavy urbanization, but for most of the year, the majority of these stream beds are dry.

Heavy water consumption by the mushrooming urban areas has caused a drop in the water table to such an extent that desert surface waters are an extreme rarity, except under the most favorable conditions. While the rains that fall on the mountains will no longer maintain surface flows in desert rivers, they do benefit vast underground streams that gradually find their way to lower levels and contribute to the substrate water tables. In a few areas, sub-surface rock formations cause waters to surface briefly, only to disappear again within a few miles into the desert sands. Clay or hardpan layers, as well as a rare abundance of water, can also bring about such desirable effects.

The surface rivers that do have a relatively constant flow have one characteristic in common: the shores are lined with a band of mixed vegetation. The plant species are referred to collectively as phreatophytes because they depend upon the moisture from the rivers' water tables. And, as a common process of nature, a certain amount of moisture is lost through transpiration. This simple fact has started a state-wide controversy and has the Corps of Engineers scrambling across the desert like so many roadrunners. The question of debate is whether or not the loss of water by transpiration warrants the elimination of the rivers' green belts.

As an excuse to despoil the last natural rivers in Arizona, the Corps adamantly maintains that the normal process of transpiration is a waste of water. Its solution to this "problem" is to strip the riverbanks of all vegetation and constantly maintain them in a manner to prevent any regrowth. In stubborn adherence to this contention, the Corps completely ignores the likelihood that the transpiration loss would be offset by increased evaporation. It also ignores the fact that stream-side vegetation is the most productive wildlife habitat in Arizona. It cannot put its usual benefit figure on the possibility of delivering a little extra water downstream, so it tends to disregard the cost of clearing and maintenance from an economic basis.

The Corps is not the only federal agency scalping the desert; the United States Geological Survey, the Bureau of Reclamation, and the Soil Conservation Service are there, also. The magnitude of the proposed phreatophyte eradication program can be best visualized by referring to the map on page 137, in conjunction with the following quotes from an article, *Statewide Summary*, by Bud Bristow, Chief Project Evaluator of the Arizona Game and Fish Department at Phoenix:

> The United States Bureau of Reclamation's Lower Colorado River clearing program, authorized by the Colorado River Basin Act, envisions the removal of all vegetation from Davis Dam to the Mexican border, except for 6,000 acres on wildlife refuges.
> The Bureau of Reclamation is also studying the San Pedro River for removal of 18,800 acres of wildlife habitat

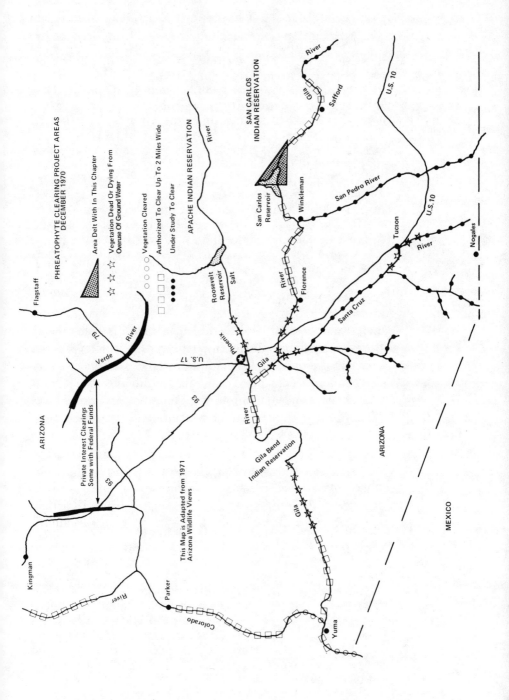

PHREATOPHYTE CLEARING PROJECT AREAS
DECEMBER 1970

▲ Area Delt With In This Chapter

☆☆ Vegetation Dead Or Dying From
Overuse Of Ground Water

○○○○ Vegetation Cleared
☆ Authorized To Clear Up To 2 Miles Wide
○○○○ Under Study To Clear

This Map is Adapted from 1971
Arizona Wildlife Views

Private Interest Clearings
Some with Federal Funds

ARIZONA

MEXICO

APACHE INDIAN RESERVATION

SAN CARLOS
INDIAN RESERVATION

Gila Bend
Indian Reservation

Flagstaff
Kingman
Parker
Yuma
Nogales
Tucson
Phoenix
Florence
Winkleman
Safford

Colorado River
Verde River
Salt River
Gila River
Gila
Santa Cruz River
San Pedro River
Roosevelt Reservoir
San Carlos Reservoir

U.S. 17
U.S. 10
79
93

from the authorized Charleston Dam, near Tombstone, downstream to the Gila River at Winkelman.

A water study involving wildlife habitat removal is presently being conducted on the Santa Cruz Basin and the Upper Gila River near Duncan.

The entire Gila River from Camelsback Dam site near Clifton to the Colorado River, a distance of over 400 river miles, is authorized for clearing except for that small area mentioned near Gila Bend. Several individual channelization and water salvage projects are included in this particular project area.

The Salt River is authorized for clearing by the Corps of Engineers from Orme Dam site near Saguaro Lake, west through Phoenix to the Gila River near Avondale.

In addition, clearing of vegetation has occurred and is occurring along the Big Sandy River, the Verde River, and several small watersheds. These particular projects are either conducted or subsidized by the Soil Conservation Service, United States Geological Survey, or the Salt River Project (Corps of Engineers).

The Arizona Game and Fish Department is charged by law with the responsibility of managing and preserving the state's wildlife resources. When federal land and water projects are planned, department biologists anticipate the effects on wildlife and report them to the responsible agencies as legally required. With regard to phreatophyte habitats, Bristow's article made the following points:

One acre of dense salt cedar will produce about as many doves as 5,000 acres of surrounding desert. Texas once had 4.5 million whitewings. The clearing of 95% of their phreatophyte dove nesting habitat along the Rio Grande River resulted in a 90% reduction in their whitewing population. They have purchased phreatophyte lands for $200-$300 per acre in an attempt to preserve the remnant population.

Arizona's clearing projects will primarily affect dove, waterfowl, deer, javelina, quail and many non-game birds species. Some non-game birds listed in the federal government's rare and endangered species report inhabit only the riparian vegetative areas of Arizona. These birds which nest in the river bottom vegetation, will no doubt cease to exist in Arizona if the current clearing plans are consummated.

The assault upon the Gila River is a prime example of the Corps' connivance in foisting its will upon the people—this time, the Indians of San Carlos Reservation.

The Gila is one of the few rivers in Arizona that is still free-flowing most of the year. Its headwaters rise in the mountains of New Mexico and wind their way across Arizona to flow into the Colorado River at Yuma. East of Phoenix, the Gila flows through the San Carlos Indian Reservation for about forty miles. This reservation was not a Utopia to begin with, but the Corps' intrusion has robbed it of the chief life-sustaining resources it did have. While the San Carlos Indians are an agricultural people, they depended upon the wild game along the river borders as a source of much-needed protein.

Game was abundant in the broad thickets of cover along the river. The Arizona Game and Fish Department reported heavy concentrations of doves—as high as 1,000 per acre. Quail, pheasant, peccaries, deer, and even a few bear were found along the banks of the Gila. Tall trees shaded much of the river, protecting it from excessive evaporation from the desert heat. This shading cooled the water sufficiently to support a generous population of small mouth bass and other fish. But the Corps, lacking in reason and compassion, stripped the land bare. The game and the fish were gone.

The Corps' procedures on the Gila ran the gamut of trickery. While it is true that some local demand for a project must be made before Congress will approve and fund it, such demand can be stimulated easily by a little lobbying. And this is a routine the Corps does not hesitate to employ, since it is one of the biggest and most influential lobbying groups in the country. It was not difficult for the Corps to get a few local people to protest the loss of agricultural waters through transpiration. This gave the Corps the excuse to launch its phreaophyte eradication program. Despite vigorous opposition from the Arizona Game and Fish Department, political expediency won out and Congress authorized the work. But for some reason, Congress had second thoughts about the project, and funding was not approved. Even this did not deter the Corps; it had visions of bigger things to come. There were more green-belted rivers in Arizona.

Being stymied for funds, the Corps turned to the Bureau of Land Reclamation which had several hundred-thousand dollars earmarked for scientific experiments. It then borrowed bulldozers from the Bureau of Indian Affairs. The Gila River plan was labeled as a "scientific experiment," and through the formation of this legally questionable triumvirate, the Corps proceeded with a project that the Congress of the United States was not willing to fund.

In a conciliatory gesture to make its own record look good, the Corps conceded to a request by the Arizona Game and Fish Department to leave a 100-foot wide strip of game cover a short distance from the river where it would have little effect on transpiration. But when the dust from the borrowed bulldozers settled, the strip was a quarter-mile from the river's edge, providing poor cover for even a jackrabbit.

Now, as the Arizona map shows, hundreds upon hundreds of miles of Arizona's rivers are scheduled to be stripped of their natural vegetation. There is something drastically wrong with the mental processes of those who promote, and of those who tolerate, such insane killing of the country's rivers.

◄§◊►   ◄§◊►   ◄§◊►

Of all the "development" atrocities imposed upon the State of California, the Corps' plan to decimate the Sacramento River Delta area is potentially the most devastating of all. This plan, vigorously supported by land and industrial speculators, is not just an ordinary bay-improvement project, but one that will cause immeasurable ecological damage to the estuaries and the bay. Also, it is a plan loaded with the possibility of catastrophic damage to human life. But the well-trained Corps has long been ingrained with the belief that human lives are expendable. It accepts this possibility and continues to promote the destruction of the Sacramento Delta.

Pennfield Jensen, editor of the now defunct environmental publication, *Clear Creek*, resides in the San Francisco area. He is quite familiar with all the vagaries of the problems and interests pertaining to the bay. In this June 1971 issue, Jensen presented a clear and concise picture of the Corps' plan and what it would mean to the bay area: With his kind permission, this brief article follows:

> San Francisco Bay, which is already more polluted than Lake Erie (according to studies by Dr. Erman Pearson, Chief Consultant for the Kaiser Bay-Delta Report), is about to become San Francisco Slough. The Army Corps of Engineers has plans to begin gutting the bay this year with a ninety-three-mile trench, from outside the Golden Gate all the way up the San Joaquin River, to Stockton (see map). This project is called the John F. Baldwin and Stockton Ship Channel or "Baldwin Channel" for short. Because Clear Creek believes every American has a stake in the ecology, beauty, and livability of the bay area, we ask you to stop this environmental atrocity.
>
> Here, for your consideration, is a cost benefit analysis of the Baldwin Channel Project:

Costs

> Water Quality: Heavy Industries—oil refineries, chemical companies, steel and paper mills—have already spent about $500 million in anticipation of the channel; and plan to spend $3 billion for heavy industrial capital if the channel goes through. These industries will draw water from the bay, and dump their wastes back in. Massive annual fish kills already occur in the bay upwaters, and the Bay Regional Water Quality Control Board says it can't guarantee that Baldwin Channel waters would be safe for recreational use.
>
> The channel would bring salt water farther into the freshwater area of the bay delta, destroying natural ecology and rendering water useless for agriculture and residents. Dredging will possibly puncture underground water basins—as happened with the dredging of the Delaware River near Trenton, New Jersey—polluting underground water with the vile channel fluids.
>
> The purpose of the channel is to open San Francisco Bay to super tankers, many of which will be loaded with oil. Tankers displacing 81,000 deadweight tons could make it to Pittsburg (Cal.); and 30,000 ton super tankers could

sail up to Stockton. These super tankers must commence "full astern" many miles before they stop, and would not be able to maneuver very much in the narrow channel.

The January 1971 oil spill was caused by collision of two tankers in the 16,000 ton class. If super tankers are allowed to penetrate to the headwaters of the bay, a truly catastropic oil spill will probably occur.

Tax Money: The Army Corps estimates the federal cost of the Baldwin Channel to be $82 million. A study of nine major projects (*The Atlantic*, April 1970) showed that costs overran initial estimates by an average of 182%. Maintenance dredging will cost United States taxpayers about $1,870,000 a year.

In addition, Contra Costa County taxpayers pay a committee of industrialists and developers—called the County Development Association—$112,000 annually to lobby for the channel and other projects. Contra Costa, a heavily industrialized county, already has the second highest tax rate in the state. The burden of state and local taxes, which are sales and property levies, fall most heavily on low and middle income residents. Channel proponents argue that industrialization would lower local tax rates. In fact, industrial growth raises tax rates. Santa Clara County and the City of Palo Alto, after studying the consequences of factories, decided it would be both cheaper and environmentally wise to halt industrialization in their areas.

Breathable Air

In February 1971, the California Environmental Quality Council told Governor Reagan that people living in the Los Angeles and San Francisco Bay areas face "a critical state of clear and present danger" from air pollution. Nine of the thirteen major bay area industrial air polluters already are located along Contra Costa County's shores. New heavy industries brought in by the channel would end all hope for breathable air around the bay.

Livable Environment

Baldwin Channel dredging would scoop up about eighty-five million cubic yards of spoils or mud, of which some sixty million cubic yards would be used to fill marshes to create land for industry. Resulting changes in

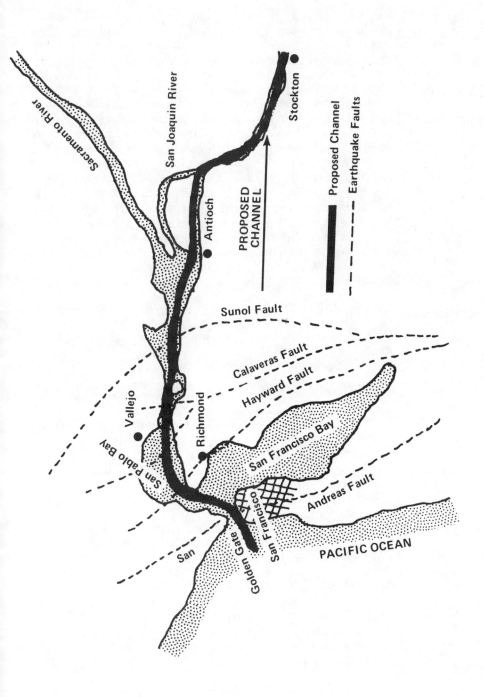

the San Joaquin River Mouth—lowered oxygen content in the water, and flow reversals—would probably wipe out the already threatened bay salmon run, the spawning route of 70% of California salmon.

The Solano County wetlands, comprising 10% of California's wetlands and providing nesting areas for about 25 percent of the ducks on the Pacific flyway, would be destroyed by fill, salinity, and factory emissions.

This whole man-made environment of channel spoils would be plastered onto a number of active earthquake fault lines. Seismologists confidently predict a major earthquake in the area. Such a quake would most likely bring softly-anchored industries down in flaming holocaust and oceans of deadly chemicals.

In short, the Baldwin Channel might allow more people to exist on the shores of San Francisco Slough; but no one would be able to live there.

Benefits

Industrial Progress: Land speculators and industrial developers, who are the major advocates of the Baldwin Channel, stand to benefit from the channel at the expense of local and federal taxpayers.

The channel is not being built to service present industry. In a poll of eight Contra Costa County industries taken by Warren Bogges, a county supervisor, only one company said its present and future needs were not met by the existing channel depth. The company that "needs" the channel is Phillips Petroleum: the number one industrial air polluter in the bay area with 90.1 tons of emissions per day in 1968, and a formidable water polluter, dumping eighteen million gallons of waste daily.

Even speculators and developers may be disappointed with channel benefits. Harvey P. Schneider, recently retired from ten years as Manager of the Federal Maritime Commission, has noted that the modern shipping industry reaps economic gain by calling at a few major terminals, such as San Francisco and Oakland. Schneider has concluded that, from an economical point of view, the San Francisco Bay region already has too many public terminals; and that the Baldwin Ship Channel is already obsolete if the consideration therefore is based upon improving the competitive position of Stockton or Sacramento.

Jobs

The Baldwin Channel would drag the bay area's economy in the direction of employing more people to produce petrochemicals, steel, pulp, agricultural chemicals, and munitions. As for the work of digging the channel itself, Senator Alan Cranston, who is pushing for early release of funds to begin the channel, notes that the first year of channel construction would create the grand total of fifty-eight new jobs.

Contra Costa Ecology Action, an active group of citizens is trying to stop the Baldwin Channel.

જ્ઉૡ્   જ્ઉૡ્   જ્ઉૡ્

The greater San Francisco Bay area, including Grizzly Bay, Grizzly Island, Suisan Bay, and Honker Bay, comprises a large ecosystem of over 200 square miles. If the Suisan Slough and the Montezuma Slough were included, the ecosystem would be double in size. This area is one enormous incubator and nursery for fish and wildlife. It is one of the principal resting and wintering grounds for birds following the Pacific Coast Flyway. Its importance as a fish resource is ecologically bound to the species of Pacific coastal waters as well as to the waters of the bay and streams. The significance of this vast ecosystem as a fish and wildlife resource is pointed out very vividly in the following excerpts from a report by the United States Department of Interior Fish and Wildlife Service:

The San Francisco Bay area comprises a major metropolitan industrial complex, including the cities of San Francisco, Richmond, and Oakland. Industrial developments extend inland and along the south shore of Suisan Bay. Sacramento-San Joaquin Delta lands are agricultural. The bay area contained 4.5 million persons in 1960. By 2020, the human population is forecast at 15.9 million.

The project would provide for enlargement of channels and construction of (new) channels, maneuvering areas, and harbors for the purpose of accommodating deeper draft vessels than those which currently use the area.

Construction would entail dredging seven channel segments, totaling 68.8 miles, inclusive of 7.5-mile cutoff along False River. Spoils from oceanic San Francisco Bar would be dumped at sea or in offshore areas via barge. Spoils from San Francisco and San Pablo would be similarly disposed of in bay waters. Materials from the remaining channel sections would be deposited by pipeline on land or in shallow water areas.

Initial spoils would total 85.5 million cubic yards. Approximately 8.7 million cubic yards would be disposed of at sea, and 16.5 in San Francisco and San Pablo Bays. Between Martinez and Stockton, about 59.3 million cubic yards would be disposed of on twenty-six land and shallow-water sites varying from fifteen to 1,150 acres. Total post-project spoils, estimated at 3.2 million cubic yards annually, would be maintenance dredged (destruction in perpetuity). The initial project construction would be five to seven years. The spoil disposal areas would be on sites used for present channel maintenance. Other spoil areas would be furnished by local interests.

Waters of the San Francisco Bay and the Sacramento-San Joaquin Delta are prime habitat for a variety of marine, anadromous, and fresh-water species. Inland bays and stream systems are of primary value as migrating routes and feeding-spawning and rearing grounds.

Principal marine fin fish found in the general offshore-coastal and inland bay areas are herring, smelt, whitebait, flounder, sandab, sole, halibut, lingcod, hake, sable fish, rockfish, saltwater perch, seabass, croaker, shark, and skate. Commercial fishing is prohibited in inland waters, but the sport fishing effort in those areas is high.

About 70% of the California salmon resource, primarily Chinook salmon, spawn in upland tributaries and headwaters of the San Joaquin and Sacramento Rivers. Migrating salmon and other anadromous fishes gain access to inland bays and streams via Golden Gate Channel. The inland migration routes follow much of the proposed project channel. The annual production of salmon spawning runs contribute significantly to the offshore-coastal sport and commercial fisheries and the inland fisheries. Annual salmon harvest attibuted to this production totals 5.8 million pounds to commercial interests, and 120,000 fish to the sports fishery. Sportsmen catch substantial numbers of steelhead trout in the delta system.

Striped bass and American shad use fresh waters of the San Francisco Bay-Delta area and upland streams for spawning and rearing. Prior to the closure of commercial fishing for striped bass in 1935, an average of about 660,000 pounds was landed annually. The present sport fishing catch averages about 1.7 million fish annually. Virtually all of the striped bass are caught in the San Francisco Delta area and in the Sacramento-San Joaquin River systems. American shad are no longer landed commercially, but prior to 1958, about 1.5 million pounds of shad were harvested annually. The sport fishery for shad in the delta and upstream waters has increased significantly since 1950. Some white sturgeon are caught incidental to angling for striped bass.

The sport fishing effort for fresh-water game fish is high. The principal species are largemouth and smallmouth bass, white and black crappies, various other sunfishes, etc.

The total sport fishing effort expended in the inland waters of the San Francisco Bay and Delta area averages about 3.3 million angler days annually.

Based on human population trends, it is expected that sport fishing effort will increase from three to five times by 2020. This estimate is predicated on the maintenance of fish resources and adequate public access and use opportunities.

The principal upland game are ring-necked pheasant, doves, California quail, and cottontails. Fur animals include some muskrat, raccoon, mink, otter, beaver, skunk, fox, and opossum.

Migratory water birds using the area for wintering and breeding include species of ducks, geese, swans, marsh birds, and shore birds.

Localities within and adjacent to the San Francisco Bay-Delta area annually support a total sport hunting effort of about 359,000 hunter days. Hunter usage could be increased considerably, if habitat and resource use conditions were augmented by effective long-range conservation programs.

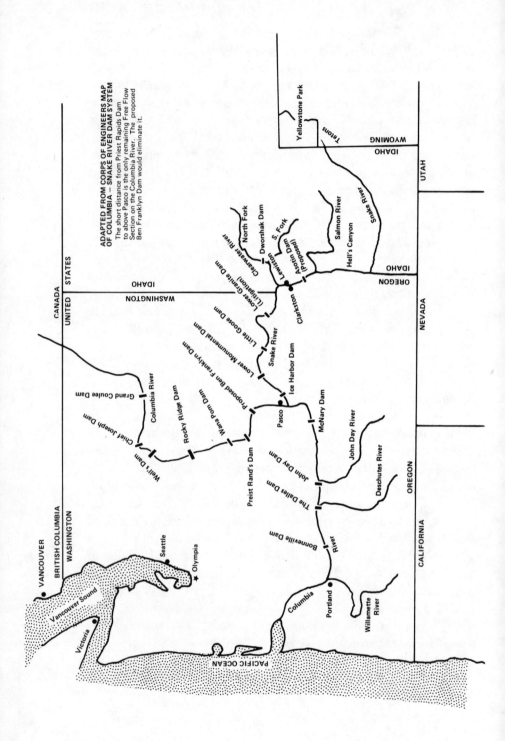

ADAPTED FROM CORPS OF ENGINEERS MAP
OF COLUMBIA – SNAKE RIVER DAM SYSTEM

The short distance from Priest Rapids Dam
to above Pasco is the only remaining Free Flow
Section on the Columbia River. The proposed
Ben Franklyn Dam would eliminate it.

There are compelling factors, other than environmental damage, that make the Corps' planned assault on San Francisco Bay completely senseless. Industrial speculators envision expansion of the petroleum industry and petrochemical plants. The bay cannot absorb the extra pollution from such installations.

On the south side of Suisan Bay, the Navy has a weapons station—a polite name for an ammunition dump. The deep, navigable water runs close to shore, and even with a limited amount of traffic, there have been reported, as well as concealed, a number of collisions and near collisions. In one near mishap, on the more open waters of the San Pablo-San Francisco Bay area, an oil tanker and a fully loaded ammunition ship narrowly averted collision. It was sufficiently close, and so potentially dangerous, that the Coast Guard held a hearing as to its cause on July 18, 1971. One such collision destroyed most of Halifax, Nova Scotia.

The Corps' plan would bring crowded navigation conditions to narrow waters. An oil or chemical fire in the vicinity of the Navy ammunition dump could result in a loss of life and property beyond comprehension. Even at this time, the several earthquake faults that traverse the region are sufficient cause for the population to fear the storage of ammunition. Why compound the danger with oil refineries, petrochemical plants, and crowded navigation conditions? The threat to the San Francisco Bay area is already strong enough without being aided and abetted by the Corps.

ᘏᔱᕒᗒ   ᘏᔱᕒᗒ   ᘏᔱᕒᗒ

To think of the Northwest is to conjure dreams of the high country—colorful mountain meadows, verdant valleys, and wild rivers filled with salmon and trout. Such dreams are not born of the imagination alone, for in reality, this is one of the most beautiful regions in all North America. It is also quite justifiably known as the "Inland Empire" because of its huge store of natural resources and the industry of its people in using the land to supply a large portion of vital products to the nation. But the creeping destruction of the Corps is here.

The great Columbia River Basin reaches from the Pacific,

across Oregon and Washington, and extends northward into Canada. The headwaters of its main tributary, the Snake River, start in Wyoming and wind their way across Idaho, through spectacular Hells Canyon, and join the Columbia near Pasco—a distance of about 1,000 miles (see map, page 148). The Blackfoot, the Raft, the Boise, and the Clearwater are but a few of the tributaries that were once magical names throughout the fishing world. But the Corps has been to this piscatorial paradise, and its dams clot the rivers like saw logs on a spring freshet. For hundreds of miles, the Columbia and the Snake no longer flow wild and free; one dam succeeds another, and their spilled waters infuse the rivers with deadly nitrogen. In 1970, more than *six million* Chinook salmon and steelhead trout were killed on these two rivers. The greatest salmon runs in the contiguous states are on the brink of extinction.

The renowned Bonneville Dam on the lower Columbia was authorized by Congress in 1933 to provide electrical power and to aid freight traffic on the river. It is here, at this dam, that salmon and trout returning from the sea, with an inherent determination to reproduce their own kind, begin the run to spawning streams some hundreds of miles inland. The run is one of tragedy after tragedy as the innate drive of these anadromous fish leads them from one deathtrap to another. The return trip to the ocean for young salmon and steelhead trout is even more hazardous.

Young fish, or smolts, returning to the sea seek fast-flowing currents. Unfortunately, this natural response to their heritage leads many of them to an untimely death. While fish ladders help some fortunate smolts around the dams, many follow the suction currents created by the powerful generators, located almost directly next to the upper fish ladder entrance, and are ground to fishmeal. And, because the free-flowing rivers have been turned into a tandem of placid lakes, food is no longer scoured up from the bottom and the shores.

But it is nitrogen poisoning that takes the greatest toll of salmonid fish, and it is the Corps' system of dams that has supersaturated the waters of the Columbia and the Snake with this deadly poison. The natural phenomenon of nitrogen in-

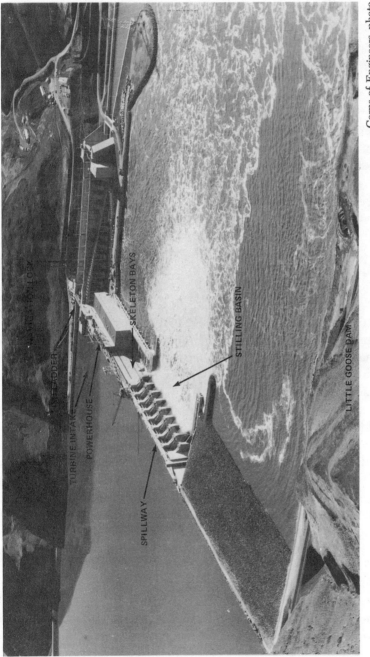

NAVIGATION LOCK

SKELETON BAYS

FEEDER

TURBINE INTAKE

POWERHOUSE

STILLING BASIN

SPILLWAY

LITTLE GOOSE DAM

Corps of Engineers photo

Of the sixteen dams built on the Snake and Columbia rivers, Little Goose Dam on the Snake is considered by many conservationists to be the worst nitrogen generator, although all contribute to the unspeakable yearly kills of salmon and steelhead. Note the proximity of the fish ladders to the turbine intake.

fusion into the waters below a high falls has been known and understood for a long time, but seemingly, this simple scientific fact escaped the engineering genius of the Corps, or else the Corps decided to ignore it.

Falling water becomes supercharged with air (largely nitrogen) and under natural conditions quickly dissipates in the relatively shallow water and fast-moving currents below the falls. The spillways on the Columbia and the Snake vary from sixty to 100 feet in height, and the torrents flowing over them fall into deep, placid pools. The nitrogen does not dissipate, and the waters become supersaturated with it.

Salmon, steelhead, and other fish breathe in this gas just as they would life-sustaining oxygen if it were there. The fish are affected much like a human diver stricken with the bends. The fish die horrible deaths. Nitrogen bubbles form under the skin, and it begins to peel away; internal organs are exploded by the gas; often, the eyes hemorrhage as they are ruptured (see photo page 155).

The Washington State Department of Ecology has stated that between 70 to 80% of the 1970 downstream migrant salmon and steelhead are killed by nitrogen poisoning. They estimate the Corps has killed as many as 2.8 million Chinook salmon and an additional 43,000 nonsalmonid fish in one migration period.

When the Corps was first confronted with these losses, it responded with the most irrational statement imaginable. It stated that only the strongest and smartest fish were surviving, thus creating a new superrace of fish—an instantaneous evolution that would take nature thousands of years to accomplish.

The success of salmon migrations is affected by still other hazards initiated by the Corps. Starting with the Bonneville Dam, fish enroute to their spawning grounds must traverse the dams by using fish ladders installed for this purpose. When fish are halted by the dam barriers, they mill about searching for ways to surmount the dams. Many of them succumb to nitrogen poisoning before they can find the relatively small fish ladders which would lead them around the dams. Fish that find their way up the Bonneville ladder are then confronted with forty miles of warm, placid water—not the cool, free-flowing, oxygen-filled waters of rapids and river currents.

After the Bonneville Dam comes the Dalles, and then the John Day and the McNary—each a major barrier to fish migrations, and each another generator of nitrogen poisoning. And then comes another dam at Priest Rapids, another at Beverly, and one at Rock Island, etc. The Columbia is no longer a river; it is a series of lakes stretched out in tandem.

The same is true of the Snake. Just a few miles upstream from its junction with the Columbia, is Ice Harbor Dam. Then comes the Lower Monumental and the Little Goose. After that comes the Lower Granite, now under construction, to be followed by the Asotin. The completion of these latter two dams depends upon the findings of the court in a suit brought against the Corps by the Steelheaders Association and the State of Washington.

The people owe a considerable debt of gratitude to this diligent and hardworking citizens group which numbers over 8,000 members. Under the excellent leadership of Arthur Solomon, Jr. of Spokane, it is a constant reminder to the Corps that the people will no longer tolerate its wanton and destructive course.

Even the scenic grandeur of Hells Canyon is threatened by the Corps and those political profiteers who support what is referred to as "the last storage project on the Columbia River system." The upper reaches of Hells Canyon have been decimated already by such dams as the Brownlee, Oxbow, and Low Hells Canyon. If the proposed High Mountain Sheep Dam is ever sanctioned, it would destroy the deepest and wildest part of the canyon. And to lose Hells Canyon would be to lose a series of land and life forms unique among the natural habitats of the world. For here one can find a transition of environment and associated life forms from desert to alpine tundra.

This succession of dams on the Columbia and the Snake disrupt the inherent time schedule of migrating salmon. In addition to the repeated delays by obstruction after obstruction, the warm, placid waters scramble the urgency and the time sequence of fish endeavoring to reach their ancestral spawning grounds. Consequently, many of the fish that manage to complete the trip in spite of all the hazards encountered, arrive too late and do not spawn. Thus, new generations of salmon are aborted. Such losses compound themselves year after year and hasten the demise of the once-great Columbia salmon runs.

The dams have also taken their toll of wildlife. The flat lands, meadows, and draws surrounding the foothills were a winter refuge for vast herds of mule and white-tailed deer, elk, and a host of fur bearers. Now, according to the state game agencies of the region, 50% of this wildlife has been lost due to the lack of suitable habitat.

The generation of electrical power was, and continues to be, the Corps' principal justification for its dam-building mania in the Northwest. The Snake River dams have raceways for the installation of six turbine generators, yet only three turbines are installed at each dam. When asked why, the Corps replies that sales of power are limited because of the tremendous oversupply from the dams already built on the Snake and Columbia Rivers. Why, then, build more dams? According to the contradictory reasoning of the Corps, the growth of the area depends on more available power.

In an area with a surfeit of potential power, this time-worn excuse has lost much of its appeal, even to members of Congress. So, in order to justify the Lower Granite Dam on the Snake, the Corps had to find a more acceptable approach. It reverted to the same ruse it used in Arkansas—the need to accommodate a lumber mill that was forgotten when the project was completed. This time, a stone quarry is the possible victim of the Corps' scheming for project approval. The Corps maintains that the quarry in particular, along with other potential industries, will benefit from the slack-water navigation that the Lower Granite Dam and the upriver Asotin Dam will provide. One wonders how the transportation of quarry stone warrants another dam on the overdammed Snake River.

There are other complications associated with this project. Reference to the area map will show that the towns of

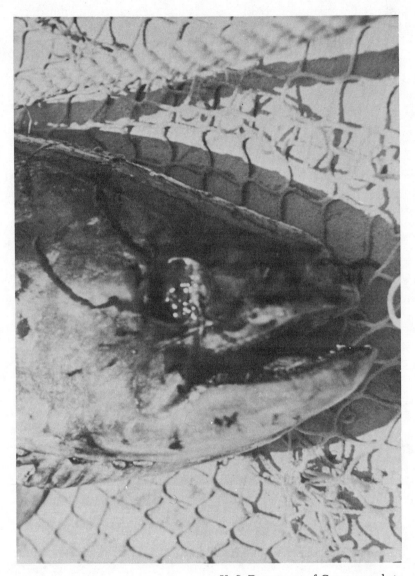
U. S. Department of Commerce photo
A nitrogen gas bubble ruptured the eye of this large Chinook Salmon. Millions have been killed this way.

Clarkston and Lewiston are located on the Snake between the Lower Granite Dam and the proposed Asotin Dam. If the dams are completed, both of these towns will be inundated unless fifteen miles of dikes are built around them. The dikes have not been approved by Congress. But the threat of rising waters will make it simple for the Corps to get approval of still another project.

Similar to the Millwood dam authorization in Arkansas that was kept under wraps for years, approval for the Lower Granite was made over twenty-five years ago. Today, the economy of the surrounding territory makes it very questionable whether slack-water navigation and unneeded electrical power justify the destruction of another segment of the Snake River.

Of all the havoc wrought upon the rivers of the Northwest, the continuing decline of the world-famous salmon and steelhead runs is paramount to citizens of the region. And understandably so, for to lose this natural phenomenon that has evolved over eons past, would be to lose another valuable segment of America's vanishing heritage. Now that the dams have been built, the problem of saving the salmon and steelhead from extinction becomes one of alleviating damaging conditions insofar as possible, and the maintenance of what natural conditions still exist.

Nitrogen poisoning is the most formidable enigma of all. What it is, how it works, and suggested solutions for its control were presented in a report by John A. Biggs, Director of Ecology, State of Washington, to the Washington Legislative Interim Committee on Fisheries, Game, and Game Fish on June 24, 1971. As a means of presenting an explicit explanation of the nitrogen saturation problems associated with the dams mentioned herein, a copy of this report follows:

Statement of the Problem

Nitrogen supersaturation is caused when excess water which cannot be passed through the generating units of the various hydroelectric projects on the Snake and Columbia Rivers during high spring flows is passed over the spillway and into a stilling basin. The purpose of the stilling basin is to dissipate the kinetic energy of the falling water as quickly as possible to preserve the structural integrity of the dam.

In the process of plunging into the stilling basin, air is trapped by the falling water and carried with it into the basin. As the water and air mixture plunges deeper, the gases contained in the air are forced into solution by the increase in pressure: thus creating air supersaturated water. When the volume of water spilled increases, the pressure also increases forcing more air into solution.

The gas primarily involved in this process is nitrogen, since in a given volume of air there is a greater percentage of nitrogen than oxygen or any other gas.

The processes, relating to spilling of water and air supersaturation have been known for sometime. Why then, is there a problem all of a sudden when dams have existed on the Columbia for many years? In the days when there were fewer dams on the Columbia and Snake Rivers, there was a sufficient number of free-flowing river reaches that dissipated the supersaturated air to a level where the effects were not critical or easily detected. The construction of each new dam made the problem more critical.

Supersaturated conditions are not dissipated rapidly in deep, quiet reservoirs; therefore, the spilling of water over one dam into the reservoir formed by another, merely magnifies the problem and it becomes cumulative.

Fish that are forced to live in supersaturated water may be affected by gas-bubble disease. This disease has been known to fisheries scientists for many years and is somewhat similar to the bends in humans. A heavily affected fish will have blisters of gas on its fins and roof of the mouth. The eyes protrude or even hemorrhage due to build up of gas behind them and death occurs. Death can also occur from gas bubbles with no visible symptoms, making the disease very difficult to diagnose. The build up of gas in the body can cause tissue damage, making the fish susceptible to other types of infection. The fish is more subject to predation due to the general weakening effect of the disease and is less tolerant of changes in water temperature. Fish that survive are more susceptible to a second exposure of nitrogen saturation.

The degree of nitrogen pollution is commonly measured by percent saturation. One hundred percent saturation is the natural condition in free flowing streams.

Under laboratory conditions, gas bubble disease symptoms have been observed in young salmon at levels of 110% saturation. One hundred percent mortality of juvenile salmon have been observed at saturation of 130%.

During 1970-71, bi-monthly samples of nitrogen supersaturation have been collected at each of the lower Snake and Columbia River projects. A review of the data during the critical period, April through June 1970, indicates that maximum nitrogen saturation values in the tailrace at Little Goose was 140%, at Ice Harbor 129% and at Bonneville 135%. The maximum values measured at these same projects during the same period of 1971 were nearly identical.

The National Marine Fisheries Service estimates that in 1970, 70% of the juvenile Chinook salmon and steelhead trout—2.8 million and 2.9 million respectively—that migrated into the Snake River from its tributary streams, died before they reached Ice Harbor and McNary Dams. These mortalities were due primarily to nitrogen supersaturation.

By comparison, in 1966 and 1967, before the completion of Lower Monument and Little Goose Dams, the survival of naturally-reared juvenile Chinook salmon from the Salmon River to Ice Harbor was nearly 100%. Significant losses of resident populations of trout, bass, crappie, and catfish have also occurred.

Solution to the Problem

The solution to the nitrogen saturation problem will be difficult, expensive, and require more time than the fishery resources can afford. Temporary measures to reduce nitrogen saturation are being utilized in an effort to save salmon resources.

In 1971, prior to and during the spring runoff and juvenile salmon migration season, fisheries agencies of Oregon, Washington, Idaho, and the Corps of Engineers worked very closely together on a day-to-day, week-to-week basis to coordinate project regulation with hatchery releases and natural migration to minimize spills and nitrogen saturation to the greatest extent possible.

For example: In order to minimize nitrogen saturation during the early release of young fall Chinook salmon from the various hatcheries on the lower Columbia, the total flow at Bonneville Dam was reduced to 180,000 cfs during the week of April 26–30, 1971, with only 40,000 cfs released over the spillway. Normal spillway flow without regulation would have been 190,000 cfs. No spill occurred at the Dalles, John Day or McNary Dams during the test period. This was made possible by storing water at the Snake River project, Chief Joseph, Grand Coulee, and Canadian projects and transferring the power load to the mid-Columbia and lower Columbia federal projects. Nitrogen samples were taken during the test and the effect of the reduced spill was apparent. The nitrogen saturation in the tailrace one-quarter mile below Bonneville, was 133% before the test, with a corresponding value of 106% in the forebay. At the end of the test, the nitrogen saturation was 103% in the forebay and 107% in the tailrace.

In addition to upstream regulation and the transfer of turbine loads certain modifications which can be made at the projects themselves are being investigated. One such modification that shows promise is the installation of slotted bulkhead gates in the upstream slots of the skeleton turbine bays. This will allow the passage of a certain portion of the excess water through the skeleton bays rather than over the spillway. The number six turbine bay at Little Goose was equipped with slotted bulkheads on May 10, 1971, and the initial evaluation of the effect on nitrogen supersaturation is encouraging. Samples collected on May 11 indicated saturation values of 106% and 104% in the forebay. In the tailrace below the spillway, it was 138%. Tests are also being conducted by the Corps relative to the effects on fish passing through the slotted bulkheads. The results should be available soon but preliminary results indicate that mortalities are minimal.

Juvenile Chinook salmon migrating down the Snake River are being trapped at Little Goose and released below Bonneville in order to shorten the time these fish are exposed to nitrogen. Presently, however, only 25 to 30% of the outmigrants are being collected.

Future project modifications which are being considered by the Corps include the installation of slotted bulkheads in the two remaining skeleton bays at Little Goose Dam and the three bays at Lower Monumental Dam prior to the 1972 spring runoff. Of course, this assumes funding will be available. The estimated cost of these installations is $1,000,000 for each project.

There will be some spilling at these projects even with the installation of slotted bulkheads. The average peak flow in the Snake River is approximately 200,000 to 250,-000 cfs. This occurs about every two years. The combined maximum flow through the power house and the skeleton bays with slotted gates installed is 130,000 cfs. On the average year 70,000 to 120,000 cfs will still have to pass over the spillway. However, the duration of the spill will be less than under present conditions.

Other Solutions Presently Being Utilized or Explored Include

1. Transfer of turbine load from noncritical to critical project.
2. Transfer of turbine load from the Willamette River and other areas to the lower Snake and Columbia River projects.
3. Transfer of surplus electrical energy to California or other consumers. These are all interim solutions and will still involve significant spilling at dams during peak flows.

Permanent lasting solutions to the nitrogen supersaturation problem in the lower Snake and Columbia Rivers is required if the valuable salmon resources are to be maintained.

As this report indicates, the Corps is making some conciliatory attempts at alleviating the nitrogen saturation problem. The use of slotted gates (first developed by the Steelheaders Association and now claimed by the Corps as its very own), the control of upstream waters, the transfer of turbine loads, and the physical transport of fish around the dams show some promise of saving substantial numbers of fish. But these are only palliative measures, for the crux of the problem lies in the simple fact that there are just too many dams on the rivers.

The Corps' overture is an attempt to assuage the public's wrath brought about by repeated engineering blunders and deceitful promotional schemes. As a point of illustration, the Dworshak Dam on the Clearwater River (a tributary of the Snake) in Idaho is a striking example. This project reveals the ignorance and the incompetency of the Corps as related to environmental matters.

The Idaho Fish and Game Department strongly opposed the building of Dworshak Dam. On March 15, 1958, it filed a comprehensive report with the Corps on a study of the effects the proposed dam would have on the Clearwater Basin. This report was a challenge that threatened the Corps' plan. The Corps countered by sending Major General E. C. Itschner (the same General Itschner that hid the Corps behind the "save and hold harmless" clauses regarding the St. Lucie Canal) to speak at a meeting in Lewiston, Idaho. General Itschner told the gathering, "If Bruces Eddy (Dworshak Dam) is going to hurt fish and wildlife resources a great deal, we would be hesitant about giving it our full-hearted support. And I have testified before Congress that there is no truth in the opposition. They are not fair, or they are completely ignorant. It would cause infinitesimal damage."

Despite public opposition, the Corps had its way, and Dworshak Dam was built. Shortly thereafter, an estimated 6,000,000 salmon and steelhead died. Now, the Corps maintains that this dam cannot be fully utilized without the building of a supplemental dam downstream! This is another example of the Corps' hit-and-miss approach to engineering.

And General Itschner was wrong, to say the least. According to a statement by the Idaho Fish and Game Department at a Corps of Engineers' public meeting concerning the Clearwater River held at Lewiston, Idaho on November 19, 1970, "approximately 50% of the anadromous fish production area in the entire Clearwater River drainage has been eliminated. The North Fork Clearwater River drainage has been completely removed from the Idaho salmon and steelhead fisheries."

This report by the Idaho Fish and Game Department is so pertinent to the problems of the Clearwater, and other Northwest rivers, that it is presented in full in Appendix G.

The Corps' propaganda machine got a big boost in 1972 when it reported the fish count at the Bonneville ladders exceeded that of recent years. Of course, the Corps quickly took advantage of this situation in hopes the public would forget about the death of 6,000,000 fish. While claiming "success" by the arrival of more fish at the first obstacle, it failed to acknowledge down river hatchery smolts as possible reasons why this was so, and it avoided the most important factor of all— the ocean survival rate. All this observation proved was that more fish endeavored to run the death gauntlet of dams, slack water, thermal pollution, nitrogen poisoning, and other obstacles put in their way by the Corps.

The National Marine Fisheries Service of the United States Department of Commerce made a more thorough study of this increased fish run. A department report by Howard Raymond in January 1973, substantiates the fact that fish mortality is still tremendous despite the propaganda and vaunted efforts of the Corps. According to the report, "survival in 1972 was no higher than 36%, and perhaps as low as 27% of the pre-dam survival rate. Preliminary evaluations indicate survival in 1970 declined to 30-40%, and in 1971 to 53-70% of the pre-dam rate. By contrast, in 1972, most of the mortality could be related to passage through slotted gates. Release of fish through the gates . . . in 1972 indicated that the mortality rate was at least 50%."

As indicated previously, the slotted gates as designed by the Steelheaders' Association showed promise of success on their initial tests. But in adopting this idea, the Corps so altered the design of the gates that they now kill as many fish as nitrogen poisoning. Yet, the Corps wants to spend $9 million for the installation of these gates!

This new cooperation being currently peddled by the Corps resembles a Trojan Horse.

Federal and State officials who have to deal with the Corps' "New Cooperation," are also constrained to tread softly, because they have to deal with the Corps on a daily basis. Consequently, public statements by them are circumspect and conciliatory. To accept this as an indication that the Corps is in fact cooperating would be a sad error.

On the Columbia and Snake Rivers the attempts to reduce nitrogen poisoning by the redistribution of electrical generation and the trucking of a few juvenile fish around the Little Goose Dam and releasing them below the Bonneville Dam, is an indication that the Corps has recognized they have gone too far, rather than any "new cooperation." They are unavoidable admissions of the frightful destruction they have caused. What the Corps hopes, of course, is that these pitiful attempts to undo a gigantic injustice will lull their detractors into a complacent state so that more and more injustices can be constructed.

If some semblance of the great salmon runs in the rivers of the Northwest is saved, it will be due largely to the intelligent and militant stands by ecological and environmental agencies of the states of Idaho, Washington, and Oregon. But they, too, must start from the basis of too many dams—dams carved into the granite canyons as permanent memorials to the Corps' wanton killing of America's rivers.

CHAPTER 8

Abolish the Corps

AMERICANS ARE TAKING a new look at themselves. In the face of a rapidly expanding population, and caught in the technological syndrome of speed, pollution, and noise, increasing numbers of people are turning to the outdoors. They are attempting to find where they have been and where they are going. They are discovering that man, after all, is not an entity unto himself above and beyond the total ecosystem, but that he is an integral part of it. There is a growing awareness of the fact that all life forms, including man, are irrevocably bound to the natural resources of the land.

Now, as Americans look back across their country, they do so with a certain degree of guilt and sadness. They find the air filled with a miasmic haze, the rich lands eroded, and the rivers stilled and polluted by the designs and wastes of man. In just three centuries, the virgin lands of a New World have been ravished by a slovenly and permissive society. And it is society as a whole that must now pay an awesome price for the greed, the profiteering, and the political aspirations that benefited so few.

In the beginning, devastation was a slow and gradual process, but it soon gained momentum with the coming of lumber barons, mining men, and agricultural interests. Their careless practices denuded the land and subjected it to erosion and flooding. This brought about the need for flood control and other protective measures—and the Corps of Engineers entered the scene.

There is no denying that some of the Corps' earlier projects were beneficial, even though the environmental costs may never be paid. It was a matter of trading a piece of the environment for hydroelectric power, navigation facilities, flood-control measures, and other benefits that could be justified on an economic basis. But the sad part of this helpful beginning was the increasing power of the Corps which continued to grow beyond all intents and purposes. Yet the Corps was not entirely to blame, for its projects were soon recognized as a means of promoting political gains. And during the depression years of the 1930's, public-work programs were wide spread throughout the country simply as a means of providing employment. During these years, and those that followed, the Corps grew stronger; economic benefits became more difficult to justify; but the American landscape continued to be ravished at an alarming rate.

The names of the rivers that have been dammed, straightened, flooded, and otherwise despoliated by the Corps are well known—Kissimmee, Oklawaha, Columbia, Snake, St. Lucie, Sangamon, Cossatot, Missouri, and so on with name after name that extols America's historic past.

In order to comprehend the awesome damages inflicted upon the country's water resources, it is necessary to view the results in total, rather than project by project. The time is long past when any sizable segment of the population can depend solely upon its adjacent environment for survival. What affects one section of the country subsequently affects all sections. And it is equally true that the water resources have a similar interrelationship. The oceans depend upon the estuaries; the estuaries depend upon the rivers; and the rivers depend upon the watersheds. Damage to one is damage to all.

The massive exploitation of the nation's waterways stands as self-indicating evidence against the Corps. The seriousness of this indictment becomes suddenly shocking when a few cumulative totals of the Corps' intrusions are revealed. In its own propaganda release of April 1973, *Historical Highlights of the United States Army Corps of Engineers*, the Corps audaciously states that "During the last 148 years, the Corps has completed more that 4,000 civil works projects. It has built more than 19,000 miles of inland and intracoastal waterways. . . . It has constructed some 350 reservoirs, and local flood-control projects incorporating more than 9,000 miles of levees and flood walls and 7,500 miles of improved channels." Of course, the Corps does not reveal the financial or environmental costs of these projects.

In compliance with the Fish and Wildlife Coordination Act of 1958, the United States Department of Interior received for review 20,000 dredge and fill permit applications from the Corps in one three-year period. Unfortunately, the Interior Department could do no more than review and make recommendations; it does not have the power of enforcement. However, the receipt of hundreds of applications per month does reveal the magnitude of the Corps' continuing assault upon the estuarine resources of the country. These totals become even more startling when the activities of the Soil Conservation Service and the Bureau of Reclamation are added.

The devastation wrought by the Corps is so extensive that every citizen within the United States is affected, no matter where he may live. The loss of a single estuary has immediate and profound effects upon local residents, but the loss of estuary after estuary along the country's coastlines is a matter of critical concern to everyone. Two-thirds of the commercially valuable ocean fishes are estuary dependent. At a time when a growing population relies more and more upon the resources of the sea as a source of food, estuary losses are a world-wide tragedy.

The results of damming and channelizing the nation's rivers have reached every segment of society. The losses of salmon, shad, and other commercial fishes are immediately recognizable, but the impact upon the country as a whole is virtually immeasurable. The great dams attract people and industry to natural flood plains who then demand physical and financial protection from their own folly. As the Corps responds to political pressures of the people, the situation is compounded and the ruination of the country's water resources becomes a vicious cycle. Environmental losses and financial burdens grow rather than diminish.

Invariably, the alteration of estuaries and waterways results in a tragic loss of wildlife, often to the point of endangering or even eliminating an entire species. The Corps and its political patrons have long been immune to such losses. Of what great importance is the loss of another species of fish, bird, or animal? This attitude fails to recognize that each species contributes to the functioning of the total ecosystem. It also fails to recognize the fact that when the system will no longer support wildlife, it will no longer support man.

The exploitation of the nation's water resources to such a far-reaching degree is a tragic reflection on a permissive citizenry. But belatedly, there is an indication of hope, even among official sources, that recognizes exploitation has gone too far, and that corrective and preventive measures are needed. Assistant Secretary of the Interior, Nathaniel P. Reed, appeared before the House Government Operations Subcommittee on June 3, 1971, and presented a statement of the Interior's views pertaining to natural-stream alterations, particularly that of channelization. His full statement is presented in Appendix F. The brief excerpts which follow are applicable to the point of immediate concern:

I was stunned last week when a copy of the Corps of Engineers' booklet *Water Resources Development by the U. S. Army Corps of Engineers in Delaware* dated January 1971, crossed my desk showing completed public works projects for that state. This report indicated there is only one major stream in Delaware that has not been channelized at least in part—the Appoquinimink River near Odessa—and the Corps has authorization to work on that.

Reviewing the status of small watershed projects in the Southeastern States alone, we found that as of August 1, 1969, 1,119 applications for watershed assistance have been received covering 122,620 square miles. Of that number, 638 have been authorized for planning and 428 have been approved for installation. Estimates indicate that projects in just this one program will involve the alteration of over 25,000 miles of stream channels to obtain flood protection and drainage objectives. These alterations will adversely affect from 25,000 to 60,000 acres of stream habitat. A conservative estimate of the wooded wildlife habitat damaged or destroyed by these alterations would be about 120,000 acres and could exceed 300,000 acres.

Studies conducted by the North Carolina Wildlife Resources Commission evaluated the effects of channelization on fish populations in eastern North Carolina streams. These studies showed that the production of game fish species was reduced by 90% following channelization. They further demonstrated that this loss is permanent because normal maintenance procedures preclude the possibility of recovery of the stream's normal productivity.

The preceding figures are just a brief glimpse of what has happened in the southeastern states. If all fifty states are included in such an analysis, the figures become astronomical. The Corps has already imposed a $10 billion burden on the American taxpayer, and it has some $16 billion more in backlog project approvals. This kind of money subjects the country's environment to alterations and devastations that defy comprehension.

The lack of public understanding has been the Corps' strongest forte in the promotion of its projects. Unquestionably, the majority of the Corps' alterations to national water resources are accomplished without the public being aware of the reasons or of the financial burdens imposed. The Corps likes it this way; an uninformed public cannot present intelligent opposition. Its plans and procedures are kept under wraps as much as possible in a deliberate attempt to present the public with an approved or accomplished fact. When suspicions are aroused and answers demanded, the Corps will resort to intensive placatory propaganda. In the case of the Cross-Florida Barge Canal, it paid a local public relations firm $1,800 a month (with taxpayers' money) in an attempt to justify the canal on an economic and environmental basis.

Although certain Corps projects in the past were ill-advised from an environmental standpoint, it could be proven that they provided needed power or flood protection. But these needs are no longer applicable to meet project proposals, and the Corps has had to adopt more devious procedures to maintain its autonomous position.

Public hearings are a prerequisite to final project approval, but the Corps must often be pressured into holding them. And, as with the Potomac River Basin project, the meetings are sometimes called at the last minute and detailed information withheld from the public prior to the meetings. When this ruse works, it prevents organized and intelligent opposition. The Corps also resorts to delaying tactics, often bringing up proposals that were approved twenty-five or more years ago. As with the Millwood Dam in Arkansas, the plans are then altered to meet "current demands"—and one dam can be parlayed into seven! And General Itschner's statement that the Dworshak Dam on the Clearwater in Idaho would do "infinitesimal damage" to the state's fisheries can be looked upon as either a deliberate attempt to hide a known truth, or as incompetency by the Corps.

It is a crime enough to trade sections of the American landscape for dollar values, but to do so without economic justification, and through the employment of deceitful procedures, is sufficient cause to question any further need for the Corps of Engineers.

Has the Corps outlived its usefulness? There are many people who think so, and with just cause. Major river systems have been dammed and dammed again; waterways have been dredged beyond practical needs; and too many estuaries have been lost. Most politically advantageous projects have been completed and pork-barrel money is more tainted than ever.

Even the Corps realizes that power needs, flood control,

and barge traffic are now weak excuses for exploiting the nation's waterways. It now strives for economic sanction by stressing dollar values for ". . . mass recreation on a large scale." It fails to recognize the fact that the recreational opportunities it destroys are often far more valuable than the boating facilities it creates. If the Corps wants to justify proposed projects by putting a dollar value on boating, it must also put a dollar value on hiking, canoeing, bird watching, stream fishing, hunting, speleology, photography, and other recreational activities that attract people to a natural environment.

Recently, proposals have been made to curtail the many powers of the Corps in the areas of independent planning and final decision making. These have come from various sources and in many different colorations. The general substance of these proposals has been to the effect that the powers of planning and decision making should be relegated to some such body as the Department of the Interior or the Department of Agriculture. In a recent message to Congress, President Nixon proposed some sweeping legislation on this matter. His proposals would embody the following general terms:

1. That the Corps be stripped of all general engineering planning and decision-making powers in setting up a project.

2. That the Corps be reduced to the status of a civilian contract engineeer.

3. That all planning and decision-making functions now held by the Corps be allotted to the Environmental Protective Agency.

4. That the engineering functions and activities of the Corps be placed under the direction of the Environmental Protective Agency.

The Corps' position of being in constant battle with nature is being challenged from other sources. Numerous proposals have been made for restricting the Corps and for assigning it new responsibilities. Congressman Henry Reuss of Wisconsin heads the House Subcommittee on Conservation and Natural Resources. He has introduced a bill (H.R. 8843) that would shift much of the Corps' manpower and resources to the construction of sewage systems and disposal plants. On the surface, this sounds like a practical move. Actually, while the bill does have more stringent regulations regarding water resource projects, it does not eliminate them from the Corps' jurisdiction. Giving the Corps the responsibility for massive sewage programs would only strengthen its position and open new channels for the flow of pork-barrel money.

Other legislative measures aimed at restricting the Corps' activities have been introduced. Senator Clifford Case of New Jersey has introduced a bill (S. 1287) that would deauthorize all unfunded projects eight years old or older. Congressman Guy Vander Jagt of Michigan has submitted a companion bill (H.R. 8754) to the House. These are restrictive rather than corrective measures, but they would at least be an initial step in limiting the Corps' built-in power.

The statement to the House Government Operations Subcommittee by Assistant Secretary of the Interior Reed included the following points which challenge the Corps' organization and usefulness:

> I think we are kidding ourselves if we do not admit that the vast majority of stream channelization has had a devastating effect upon our nation's waterways. We could spend all day detailing the endless miles of streams slated for additional modification by one agency or another. But that will not solve an admittedly serious problem. What is needed is a complete rethink and redirection by the men who are designing and constructing the projects.
>
> Even though we spend millions of dollars each year for ditching, dams, and diking of our rivers and streams, the flood damage throughout the nation continues to rise. Perhaps our philosophy has been misdirected. We have some federal agencies charged with doing a job which involves environmental destruction and others charged to protect the environment in continuous conflict.
>
> . . . it has been the observation of the majority of our personnel that those agencies engaged in stream channelization activities are still largely paying nothing more than lip service to earnest environmental protection. We have yet to detect any substantive departure from the practices of yesteryear by these agencies, and I believe the record will clearly support these conclusions. . . .

A most significant evaluation of the Corps' independent position, and recommendations for corrective measures, were made by Dr. Bruce M. Hannon, Assistant Professor of Engineering at the University of Illinois, as a part of the report of the Committee on Allerton Park (see Chapter 6). This report by Dr. Hannon is a brilliant analysis of the entire Corps problem, and its inclusions for solutions are undoubtedly the most practical of all those presented to date. For this reason, and with Dr. Hannon's permission, the report follows.

The Anomalous U. S. Army Corps of Engineers:
A Long Needed Revision
By Bruce Hannon, Ph.D.

More than one professional person has remarked that the dam building and canal building antics of the Corps of Engineers is innocent stupidity elevated to an art. Still other critics think that the Corps is vindictively pursuing an environmentally degrading course (molding the environment for man) in order to maintain its image and position as the world's largest construction agency, consisting of 32,000 civilians and 220 officers.

Basically, the Corps carries out policy set by the Federal Water Resources Council (FWRC) a group whose board is composed of members from the federal water agencies. By judiciously locating retired Corps staff on the council, however, the Corps has a remarkable influence over federal water policy. For instance, the council's second in command, Reuben Johnson, is a retired Corps Colonel. The Corps also performs the planning, justification, assists Congress in authorizing and financing, and does the contracting for their $1.1 billion in annual water projects expenditures. The top Corps officials control a completely contained process which fortunately is incurring public wrath over environmental damage and financial waste.

The following seven suggestions are offered as ideas to help remake the Corps into a socially useful bureau without the undue trauma of complete eradication. They apply just as well to the other mission!-oriented anomalous federal agencies such as the Bureau of Reclamation and the well-derailed Soil Conservation Service.

Suggestion No. 1: No member or former member of

any federal water bureau should be allowed to serve in a water-policy-making role. This eliminates the built-in positive feedback loop which the Corps has constructed in the FWRC. The Corps also has officers and civilian personnel in such prestigious groups as the Council on Environmental Quality and the office of Science and Technology.

Suggestion No. 2: A realistic time-in-grade limit should be set on those civilians in the upper echelons just as in the armed services. This "up or retirement" criteria would keep the Corps' civilian population competitive and would tend to single out the best decision makers. A twenty-year maximum service limit should be placed on Corps personnel in upper echelons allowing the best to be retired or act as consultants after twenty years. It is very likely a least-social-cost process to retire with pay those Corps civilian personnel who cannot meet specific quality standards, rather than let them continue to plan and make multi-million dollar mistakes, such as the $150 million Cross-Florida Barge Canal or the $65 million Oakley Reservoir Project in central Illinois.

Suggestion No. 3: Separate the Corps from the Army, and rename it the Federal Engineering Administration. The Corps' connection with the Army seems to enhance military funding by exchanging public works projects for DOD support. It also provides a few hundred officers who serve as a public relations facade. The officers are often concerned only about their efficiency reports during their short (three year) tour with the Corps. The average officer in the Corps is not troubled or even aware of the errant course of his temporary bureaucratic home. To speak out, he fears, would mean trouble at promotion time.

The Task Force on Water Resources and Power (Hoover Commission Report—June 1956) noted that the Corps wished to retain its officer cadre since it prepared these men for combat-construction zone experience. As it turns out, the main combat training the Corps officers receive these days is at the hands of local women's or garden clubs.

Suggestion No. 4: The Federal government should develop a national land-use policy and planning agency. Such a device has been recently suggested by President Nixon. The planning function which now exists in the Corps should be removed to this new bureau. The Corps' amazing ability to justify or unjustify a project keeps many a pork-hungry congressman and chamber of commerce in line. Likewise, powerful congressmen such as the late Senator Kerr or Congressman Kirwan, could command the Corps to justify projects for certain areas. A look at the distribution of Corps projects on a map of the National Waterways illustrates this point. At first glance it would seem that Oklahoma, Arkansas and Ohio contained all the U. S. rivers, but in reality they are or were the domains of powerful public works-oriented U. S. Congressmen.

A separate planning agency would be less sensitive to such apparent pressures. Of course, any new planning agency must have its plans subjected to review by other federal agencies such as the OST (Office of Science and Technology) and CEQ (Council on Environmental Quality).

One of the greatest defects of the Corps is its lack of consideration of alternatives. The Corps promotes structural flood control (dams and channeling) over flood plain management. It sees dams as the panacea to national water supply and recreation needs. It continues to promote hydroelectric power and above all continues to provide free waterways for the barge industry. Twenty-six percent (about $325 million) of the Corps' budget for fiscal year 1971 went into planning, construction, operation and maintenance of barge waterways. If barge lines were not subsidized by the Corps, they would have to compete with the nation's troubled railroads on a more even basis. The Corps has simply selected a list of public needs which it thinks can be satisfied by dam construction. The Corps usually responds to such criticism by saying that it is legislatively limited and keyed to its present course. Given the Corps' seemingly inextricable connection with Congress, it seems unlikely that legislation stands in their way. It actually appears that the Corps lacks bureaucratic responsibility and wishes to continue its comfortable but irrelevant course until the last tributary to the nation's last stream is stagnated.

At present, flood plain development is encouraged by unjustified confidence in the ability of Corps' dams to prevent all flooding. Very large subsequent floods create enormous damage, causing increasingly high national flood damage despite the fact that the Corps and similar federal bureaus have dammed almost all of the country's rivers. Funding for dams and canals should be suspended until uniform federal flood-plain zoning standards are adopted.

The Corps fails to recognize the two major criteria of planning: planning without specific population limits is not planning but promoting, and the best plan is one which maximizes the number of future possibilities (i.e. public ownership of river bottomlands allows a future return to farming, damming or continuance as a natural stream, while damming and channeling are terminal commitments.)

Suggestion No. 5: Transfer the Corps' project justification process to the Office of Management and Budget. Separation of the power of economic justification from the Corps is the major step of these proposed changes. For the same reasons that the planning function was moved to a new agency, the economic justification process must be removed to still a third location in the executive branch of government. The reasons are numerous. The Corps was found in the Hoover Commission Report to overrun their estimated cost on an average of 30%, accompanied by severe overstatements of benefits in apparent attempts to justify projects which were politically feasible.

Early authorization of projects is constantly accomplished by the Corps in an effort to fix project interest rates at the lowest possible level. The authorization is often premature as indicated by comparing final projects with their authorized plans. The Corps claims that minor project refinements are within its discretionary authority. For instance, the Corps added an extra reservoir, doubled the cost and quadrupled the land required for its Oakley Project in Illinois and called these changes "minor refinements."

During periods of budget cuts, the Corps anticipates by artificially enlarging its budget requests. Resulting cuts leave it generally with a budget at least as large as the previous year. When the budget cuts are deeper, the Corps slows or stops funding of projects which are well under way. It spends the reduced funds on new construction starts and land buying for new projects in order to maintain an increasing level of total projects expenditure. The OMB is relatively free of such bureaucratic skullduggery.

Suggestion No. 6: The authorization and financing procedures followed by the Corps are, in my experience, a sham. "Local public hearings" really mean "hearings for local supporters" who are well notified in advance. The ABC Valley Coordinating Committee means a group composed of Corps and SCS representatives with a small entourage of recreation, wildlife and transportation specialists, led by the Corps. The Corpsman apparently in charge is the colonel who is Engineer for the District but a closer look reveals the real leader, an elderly civilian who likely joined the Corps during its depression heyday, the 1930's.

The every-election-year public works omnibus coasts thru the House and the Senate, reeking with approximately forty pork barrel projects on the average. Each project has passed the necessary test of satisfying at least one Congressman, one Corps district office and one chamber of commerce. The President himself cannot blast a project out of a bill; he can only veto the whole bill. Only two Presidents have ever dared.

Each congressman who might oppose this rank waste of the public funds ($1.1 billion annually to the Corps alone) simply will not speak out through fear of retaliation on some future bill of his own. Only a few such as Senator Proxmire or Senator McGovern have the courage to attack appropriations for the Corps' public works projects. As an example of the force of the congressman's will, Senator Cooper openly opposed the Corps' Red River Gorge Project in Kentucky to no avail. Finally, through Senator Cooper's urgings, President Nixon decreed the Gorge must be preserved. Even this unusual procedure only moved the proposed dam sit a few miles downstream of the Gorge and the original project may well return with the coming of a new president.

The Corps has a literal stranglehold on the authorization process. The infamous Rivers and Harbors Board contains Corps members and is composed entirely of those who believe that what is good for the chamber of commerce is good for the country. This board, which appears to possess microscopic sensitivity for the natural environment, would be dissolved in lieu of the planning agency in *Suggestion 4* above.

The concept of the omnibus bill should be removed to the history books as a failure in legislative administration. Each public works project should be authorized separately and with minimal discretionary authority.

The appropriations committee's public hearings are largely a stage production by only one very bored-acting Senator or Congressman. The hearing officials alone cannot be expected to digest complex technical, economic, social and environmental arguments against public works

projects which they have been managing since the Great
Depression. One committee staff member and his Congressman will approve billions of dollars annually in public works projects strictly on the basis of testimony of
agencies such as the Corps. This is certainly an economically and environmentally dangerous procedure.

The technically competent citizen who appeals to the
appropriation committee quickly concludes that the de-
cision-makers do not weigh technical input. The process is
purely political and replete with appropriations trade-offs.
Public Works appropriations hearings seem to the citizen
environmentalist to be the forum where political debts are
paid in concrete; the tally line of a sort of huge Congressional debit and credit page, settled annually with the
passing of pork barrel appropriations. This is of course an
environmentally degrading process or it would never have
been questioned so often as it has been these past few
years. The only possibility of stopping it is to turn the
public spotlight on the process forcing the year-end settlement to occur in the media. The solution lies in the
adaptability of Congress to new social needs and its willingness to give up these outdated means of patronizing
its fellow members.

Suggestion No. 7: The remaining functions of the
Corps now separated from the Department of Defense
and stripped of its planning and economic justification
process are designing and contracting. If highway, park
and other functions were pooled with the Corps' waterway
design expertise, a federal engineering bureau could be
formed. The additional functions of contracting, super-
vision, inspection and operation and maintenance fit naturally into the concept of this new bureau.

Revisions of the type suggested here will not come
easily. Congressional legislation could handle the problem, but Congress has shown no willingness to tamper
with the Corps' status. The insufficiency seems to occur
as a result of the lack of communication between the Congress and the electorate. As a Congressman rises in senior-
ity, he comes to rely increasingly on information brought
to his Washington office. The countless bits of information
which form the congressional commitment are actually
supplied by lobbyists, representatives of vested interests.
Few older Congressman recognize the need for real information gathering meetings with the informed people of
their district or state.

The major effort needed on the part of Congress is a rebuilding of the current Public Works Committee structure to suit the proposed separation of functions. Congressman John Saylor of Pennsylvania has long urged such restructuring of that committee.

Congressman Henry Reuss has suggested redirecting the Corps by charging them with the responsibility of constructing sewage-treatment plants. The Corps has taken the suggestion seriously and has started planning 1,000 square mile sewage and solid waste disposal area in down-state Illinois for the city of Chicago. The Corps is considering the alternate of pumping the sewage through an underground aquifer to the disposal area. Besides being an expensive project of questionable land use, the concept tends to convince Chicago residents that there is truly a mystical "away" where their wastes can go without negative feedback. This kind of planning is obviously promotive in that it allows unfettered increases in per capita waste disposal and ultimately, increases in population.

The major issue in redirecting the Corps into sewerage, however, is the disposition of the 32,000 man civilian staff who are trained to design and promote the damming and channeling of streams. These people are unacquainted with sewerage; therefore the new dimension proposed by Reuss means a bigger Corps, with continuing destruction of our water ways. Unless we disemploy them, we must remember that governmental decisions required Boeing Aircraft to terminate employment of 60,000 people inside of three years. They've set the pattern.

There is a new voice in the land. It makes headlines for the news media; it speaks with authority in the courts; and it resounds through the halls and offices of Congress with penetrating annoyance. It is the voice of environmentalists—the voice of an awakening people. Even if belatedly so, Americans are recognizing the values of their natural environment. They are becoming involved, and their voices are being heard.

It is increasingly difficult for the Corps, and for Congress, to quietly pursue a project and then present the public with a staggering bill for a *fait accompli*. When Congress passed the National Environmental Policy Act (NEPA), it did so with a substantial margin. Perhaps it was overly anxious to get on the environmental band wagon, but in doing so, it inadvertently gave project opponents a strong legal tool in the required impact statement. Many projects now in litigation have halted the flow of promised pork-barrel money. A number of Congressmen are now disenchanted with their own directives, and there is considerable agitation to change them. One of the proponents eager to change NEPA is Senator James O. Eastland. When his favorite Tennessee-Tombigbee Waterway project was halted in court, he complained, "Everybody is in favor of protecting the environment, but this business of yelling 'ecology' every time we get ready for a new project has got to stop." This is the same Senator Eastland from sunny Sunflower County, Mississippi, who in 1968, received $166,978 in federal farm subsidies for not farming existing lands! What, then, is the basis for encouraging projects that will create more agricultural lands?

Whether NEPA survives intact or not, it has already established the fact that the public's voice must be listened to with respect. The word "environmentalist" is no longer just a label for new hard-core visionaries; it is applicable to citizens everywhere. This concerted understanding and concern becomes a formidable political weapon against those who would exploit the country's resources for personal gain.

Former Secretary of the Interior, Stewart Udall, recognized the potential influence of this new group and some of the problems it must face in a copyrighted article of June 1971. With Mr. Udall's permission, that article follows.

Stewart Udall
ARMY ENGINEERS TRAPPED BY PAST

Few bureaucracies are more rigidly trapped by their own past than the U.S. Army Corps of Engineers. Resembling brachiosaurus, a giant waterloving dinosaur with less brains per pound of flesh than any other vertebrate, the corps has survived from the Jurassic Age of Engineering when dams, locks and dredged-out channels were deemed man's finest gifts to nature.

Today the corps sits atop a $16 billion backlog of public works projects inherited from this primitive age—but it must dust them off at a time when nearly everyone is an environmentalist. An outraged public has risen up to challenge the corps' antediluvian cost-benefit ratios on the Tennessee-Tombigbee, the Kickapoo, the Cossatot, the Sabine-Trinity, the Columbia, the Cheyenne, the Sangamon and scores of other rivers whose very lives are at stake.

Conservation Foundation president Sydney Howe spoke for millions of Americans early this year: "Few, if any, of our institutions face a greater crisis of confidence than does the public works establishment."

Slowly and defensively the corps has responded to this crisis. In April, 1970, Gen. Frederick Clarke, the chief of engineers, appointed a six-man Environmental Advisory Board of outside experts. The board was prematurely dismissed as "window dressing" by a few prominent conservationists, but under Charles H. Stoddard, former director of the Bureau of Land Management, it has earnestly tried to uplift the Jurassic mentality of the corps.

Now, more than a year after its birth, the board has precious little to show for its efforts. It has won the ear of Gen. Clarke, but he either cannot or will not halt the foot-dragging of 38 district offices that enjoy the ironclad support of Congress and therefore considerable autonomy from corps headquarters.

Having watched the work of the Environmental Advisory Board since its debut, we can report that very few of its recommendations have been followed. For example:

—The board asked Clarke to issue tough, precise environmental guidelines to his district engineers. Result: On Nov. 30, 1970, Clarke released a powder-puff document that the board refused to endorse. In February, however, Clarke wrote Rep. Henry Reuss (D-Wis.) a letter implying that the new guidelines had the board's support.

—The board called for a greater public voice in the early planning of new projects. Result: After eight months the corps has not even acknowledged receiving this request. It has not built in a more rigorous public review of its waterways projects.

This kind of treatment has begun to disillusion some of the board members, though not all to the same degree. "At the highest level the response has been good," says Lynton Caldwell, the board's resident political scientist from the University of Indiana.

"However, Clarke faces powerful countervailing forces. His own district staffs and their allies in Congress still approach pork-barrel projects with a 19th-century spread-the-wealth philosophy."

Stoddard, the first-year chairman, may soon quit the board in disgust. "Progress has been glacial," he says. "What the corps needs is a completely independent agency, with veto powers, to review all its public works. For starters the entire $16 billion backlog must be de-authorized and looked at fresh, project by project."

The Environmental Advisory Board has tested the willingness of a bureaucratic dinosaur to adjust to a new age. The results are discouraging. Says Stoddard: "Our experience proves that the corps' ability to reform itself from within is very, very minimal. The big changes must be imposed from without."

In short, it's up to Congress—which created the public works monster in the first place.

The Corps has already indicted itself. This new political body must recognize this fact and demand corrective procedures. These procedures must also reach into the halls of Congress, for this is the only place where the continuing assault upon the nation's waterways can be stopped. The Corps and Congress have been so accommodating to each other that radical changes will be difficult to accomplish—but the power of the vote can do it.

This new body politic will find that past performances by Congress have been a sham. The "understanding" between Congress and the Corps concerning water resource projects was so strong that committee hearings, authorizations, and appropriations were merely taken-for-granted procedures that old guard Congressmen contended with in a most nonchalant manner. For example, when freshman Congressman Pete du Pont of Delaware appeared before the House Public Works Appropriations Subcommittee regarding the proposed Tocks Island Dam, his allocated time of thirty minutes was cut to five minutes. He barely had time to introduce the subject and his supporting experts from the University of Delaware until Chairman Joe Evins of Tennessee stopped all testimony. When

Jack Paxton, representing the Committee on Allerton Park, appeared before the Senate Appropriations Subcommittee, only one out of seventeen committee members was present to hear his testimony. And the Chairman, Senator Allan Ellender of Louisiana, told Mr. Paxton that he had asked all his questions about Allerton Park *last year!*

The Corps uses these Congressional Subcommittees to further its own cause and the political wishes of Congress. All projects up for appropriations approval are lumped together in one omnibus package. The Corps is careful to see that the package includes projects in the proper states. Thus, the omnibus bill is assured of passage, because committee members will not vote against their own pork-barrel allotment. In turn, Congress, placates the Corps. It gives the Corps blanket approval for all projects costing less than $1 million, and it can spend as much as $10 million with only committee approval. It is this kind of chicanery that needs corrective measures.

As we view the carnage left in the path of the Corps' persistent and devastating march across the land, we can but wonder: "How much more can America stand? What push of a bulldozer, or what swipe of a dragline will trigger ecological disaster for our country?"

This concern is real indeed, because the Corps has plans for the future that will increase the momentum of destruction and overwhelm its ditch-digging antics of the past. By 1980, the Corps wants to double the capacity of the nation's reservoirs; it wants to double the 1960 capacity of flood-control works—from 219 impoundments to 738; and it has plans to add another 7,000 miles of levees and flood walls, and over 3,000 miles of channel "improvements."

The Corps is also paying homage to other schemes so awesome in nature that future generations will be subject to water-management fallacies of a vanished society. The ecological damage of a proposed $200 billion North American Water and Power Alliance, a plan that would divert Canadian Arctic waters to the United States and Mexico, is so frightening as to be beyond human comprehension. And the multibillion-dollar plans for Texas and California are also filled with ecological horrors.

With such designs on America's future, there are certain questions that must be answered *now*: Do we want to continue paying for projects that are not wanted or needed? Should we continue to subsidize a barge industry that ruins river after river? Should we support hydroelectric projects in Hells Canyon and elsewhere throughout the Northwest where additional facilities are not needed? Should we support hydroelectric projects at all, as they are being made obsolete with increasing numbers of nuclear plants? Do we want to invite complete ecologic disaster? Should we continue to tolerate the Corps and the politicians who placate it? If the answer to these questions is not a resounding "No," then America's future lies in extreme jeopardy.

As noted previously, there are those who would restrict the Corps' activities, or shift its responsibilities elsewhere. But are these not just conciliatory or palliative measures at best? With the current alliance between the Corps and Congress, pork-barrel money will continue to flow and the environment will still be the victim, whether in the rivers or elsewhere. For the growing masses concerned with saving outdoor America, there is only one positive answer: completely abolish the Civil Works Branch of the Army Corps of Engineers!

Ironically, it was Congress who created this public works monster, and it is Congress who must get rid of it. There is little doubt that Congress will be reluctant to do so, but a united people with a common cause of saving America can convince permissive legislators that they have no other alternative. It is time to inform Congress that they tell these ditch diggers to pack up their machinery and go back to military duties, or we are going to put someone in their place who will. After all, it is not a few pork-tainted campaign contributions that keep them in office; it is the total number of votes that express the wishes of the majority.

America has no other choice.

There was once a beautiful, fertile, green land endowed by nature with good earth and plentiful resources. Then man scarred it with a checkerboard of irrigational ditches and canals. Where great cities and dense population once flourished in ancient Mesopotamia, a present-day Iraq stands as a country of desert sands dotted with historical ruins, a monstrous monument to those who would drain the land dry. With so much evidence on our doorstep it is hardly necessary to look elsewhere. But the renowned Aswan Dam on the Nile, in just a few short years has reduced the once lush agricultural economy of the Nile Delta to a struggle with the desert—which threatens to reclaim its own, has acquired for the area a non-supportable economic reliance on commercial fertilizers—to replace rich soil formerly provided by nature for thousands of years, and has caused a ruined Mediterranean estuary—where once abundant sea life no longer exists.

These tragedies are but symbolic of what this book shows is already happening in America, and why. Will this fate be ours? Must we tolerate

The River Killers?

Appendix A

FRESH WATER IN ST. LUCIE ESTUARY—
GOOD OR BAD?

Prepared by U. S. Army Engineer District, Jacksonville Corps of Engineers.

What happens to fish and fishing when large quantities of fresh water are released into a saline estuary? Specifically, what are those effects at St. Lucie estuary near Stuart on the lower east coast of Florida when regulatory discharge from Lake Okeechobee becomes necessary? Those are questions which the U. S. Army Corps of Engineers, in charge of lake regulation, have long asked themselves.

The same questions and many other related ones have also haunted some of the people near Stuart, where statements have been broadly publicized that fresh-water releases from the big lake were synonomous with the end of fishing in their estuary. This word got around, and quite naturally the tourist business dropped off. But are the overall results good or bad for the fishing? Many people had opinions, but few really knew the answer. So much biology was

185

involved that layman opinion—experienced though it may have been—was inadequate to supply the answers.

The Army Engineers looked at it this way. St. Lucie Canal was originally dug between 1916 and 1924 by the State's Everglades Drainage District to provide the principal outlet of Lake Okeechobee. The Corps toik it over in 1930 after the disasters of the 1926 and 1928 hurricanes had drowned over 2,000 people in the Lake Okeechobee area. The Corps has enlarged the outlet and depends on it for regulating the lake. Of course, large substantial levees have been built by the Corps along the lake shores, but eventual release of that stored water is at times essential. Sometimes, maximum release by way of every possible outlet must be effected to preserve human safety in the lake region. Good or bad, there is no other way but to discharge by way of the St. Lucie Canal when the lake gets sufficiently high.

Certain residents and businessmen in the estuary area, however, took a dim view of this fresh-water discharge. From various sources, the following categorical statements were made:

1. Our famous sailfishing out in the ocean is adversely affected.

2. Small fishes and sometimes larger ones are killed in the estuary by fresh-water discharges.

3. Marine game and sports fishes leave the area.

4. Sports fishes are dispersed throughout the estuary so as to be hard to catch.

5. Sports fishes will not bite when the water is turbid.

6. Commercial fishing inside and outside the estuary is damaged.

7. Crabs and shrimp are driven from the area.

8. The effects on fish and other organisms, and their habitats, continue long after the discharges are stopped.

Were these claims valid? The Corps had to know. The time for guessing and opinions was over. The Engineers had their own biologist, Gordon E. Hall, who studied the problem at some length. They also called in as a consultant, Dr. Gordon Gunter, Director of the Gulf Coast Research Laboratory and a widely renowned bioligist. The two biologists worked up a program to get the answers over a 3-year period by scientific testing and sampling. Biological samplings with trawl nets and seines were coordinated with times of variable discharges from the lake, ranging from zero flow to maximum discharge. Investigations of water turbidity were also correlated with the biological samplings. The ability of the estuary to return to high salinity after a long discharge period of fresh water was taken into account. Every phase of the problem would be evaluated and investigated. The chips would fall where they may.

Dr. Gunter was selected for this job because for many years he has been an authority on the relation and distributions of marine organisms to the salinity of their environment. He is a former marine biologist of the Texas Game, Fish and Oyster Commission; a former director of the Institute of Marine Science of the University of Texas; and a special investigator for the U. S. Bureau of Fisheries. He also served for a year as Professor of Zoology in the Marine Laboratory of the University of Miami and for another year as Senior Marine Biologist at the Scripps Institution of Oceanography. Hall is a former director of the Fisheries Division of the Oklahoma Department of Fish and Game. Gunter began by stating that he was more interested in the theoretical considerations of the work than the problems of the Engineers, and was more interested in finding out what species lived where at various salinities than in the local problems. Gunter and Hall are now in the process of writing the technical reports on their findings. Meanwhile, the following summary concerns certain facts which have a bearing on some of the statements listed above.

Ten rounds of stations in the north and south forks of St. Lucie River and in the outer estuary were made between January, 1957 and January, 1959. Lake discharge into the St. Lucie during five of these periods was zero. During the other five periods, the discharge ranged from 2,160 cubic feet a second to 7,380 cubic feet a second. These various field surveys covered all seasons, and, as it happened, one of them was made during and following some of the coldest weather ever experienced in the Stuart area so far as Weather Bureau records go.

Trawls and seines were used for sampling of small and young fishes and animals in the estuary because (1) the small forms are more easily and reliably captured than adults, (2) their abundance is tied in with and results from the productivity if the area, and (3) comparison of the numbers of these indicator species at the same stations during various discharge conditions would serve to indicate the bioligical effects of fresh-water releases.

Slightly less than 25,000 specimens of fishes were caught. Sixty-four species of marine fishes and 24,151 specimens were taken; 19 species of fresh-water fishes and 632 specimens were caught. Over 18,000 marine fishes were taken where the water was less than 0.5

part per thousand saline. Most of these were small fishes and over 10,000 of them were baby mullet. Menhaden, croakers, and silversides were the other most abundant species. Slightly over 4,800 fishes were taken when the gates were closed and slightly less than 20,000 specimens were taken when the gates were open. Sixteen common marine fishes were taken where the salinity was less than 0.5 part per thousand, which is virtually fresh water. The salinity of sea water is 35 parts per thousand. The commonest fishes taken were striped mullet, menhaden, croaker, silversides, bay anchovies, and mojarras. These species comprise 90 percent or about 22,000 of the total catch. Blue crabs, brown shrimp, and white shrimp were found in salinities less than 2 parts per thousand. These findings bear out what is already known concerning the distribution, life history, and growth of the various fishes and invertebrates involved.

During January and February 1958, large numbers of fishes were killed by the cold. These fishes were killed in the north and south forks and in the lower estuary in conditions ranging from fresh water to quite high salinities. The fishes most commonly killed were tenpounders and large mojarras (sand perch). Large mullet, tarpon, and snook were also killed. Small dead fish were absent, which is the usual occurrence with fish kills by cold. These kills were not limited to the St. Lucie estuary, since during the same period fish kills from cold weather were reported in coastal waters from South Carolina to Texas.

Information obtained from the U. S. Fish and Wildlife Service and the University of Miami Marine Laboratory during the time the investigation was carried out shows that the volume of commercial fish catches in the St. Lucie area did not decrease with the fresh water, although certain species were taken more, and others less, abundantly in discharge periods. Also, records show that sports fishing licenses in the Martin County area have increased several hundred percent during the past 10 years. It was also found that fishing in and near the St. Lucie lock increased greatly. The Corps of Engineers had to build special guard rails for the hundreds of people who came to the locks to fish. Snook were the most abundant marine fish caught. Large numbers of mullet, catfish, bluegill, and other species were also taken.

Salinity surveys made during the course of the biological work showed that a tongue of salt water always exists in the lower estuary even during the highest flows through St. Lucie Lock (7,400 cubic feet a second.). The survey showed that pompano, pipefishes, lookdown, moonfish, bumper, sardines, boxfish, and spadefish were absent when fresh water was being released. Although no bluefish or mackerel were collected in the seines or trawls, these species, which generally stay in the higher salinity waters near the inlet, undoubtedly moved out during high discharge periods, and fishermen reports support that belief. Inasmuch as menhaden, mullet, and anchovies are forage fishes (or food fishes for others), the biologists concluded that a moderate flow of fresh water into the St. Lucie, which promotes the growth and reproduction of these species, was better than no flow at all and also better than a high flow.

Various nutrients in the form of nitrogen compounds and other organic materials are brought down from the lake. The energy processes of the earth and waters are such that these materials are inevitably used and probably significantly influence the large amount of forage fishes produced when moderate flows at the locks are taking place. The biologists have recommended that moderate discharges—up to 2,500 cubic feet a second—be made at St. Lucie Lock during January and February when mullet and menhaden are growing rapidly. They feel that this will increase growth and survival of forage fishes needed by other fishes, also young game fish and commercial species, and that it will enhance the general productivity of the St. Lucie estuary. This is an interpretation based on indubitable data. Other interpretations can be placed on the findings, but these data themselves are unquestionable. The biologists also recommend small discharges during June and July for improvement of snook fishing below the locks and in the inner estuary. This is based on fishermen use and catches during all types of outflow from zero upward. The above allegations on the effect of fresh water on the fish life of the St. Lucie may be examined in the light of these findings and other well-established biological knowledge.

It has been claimed that sailfishing off the coast has been seriously affected by fresh water from the St. Lucie. Investigators from the Marine Laboratory of the University of Miami have pointed out that this statement has no validity because sailfish are denizens of the high seas and the flow of waters from the St. Lucie has no effect upon the salinities of the Atlantic Ocean waters where sailfish are caught.

There is also no evidence that small fishes and sometimes larger ones are killed by fresh water. There is no doubt that certain marine game and sports fishes leave the area when fresh water comes. So far as the present evidence goes, this concerns two desirable sports species—the bluefish and the pompano. Sport fishes collected in the estuary during periods of high flows and low salinities included croakers, spot, sand perch, catfishes, snook, white trout, ladyfish, jacks, gray snapper, yellowtail, whiting, black drum, sheepshead, and pinfish. There is no evidence that marine sports fishes disperse throughout the estuary and, in fact, the snook certainly seem to concentrate around St. Lucie Lock when fresh water is being let out. Thousands of people use these areas for fishing. There is no evidence from the above-mentioned investigation that sports fishes will not take the bait as well when waters are turbid. It has been shown, however, that fishing for snook is quite successful when the waters are brown and turbid. According to statistics of the Fish and Wildlife Service, commercial fishing has not been harmed in the St. Lucie. Salinity determinations show that the effects of fresh water are transient; so transient, in fact, that fresh-water inflows during the fall do not lead to low salinities during the winter. The high-salinity conditions and the marine fishes that were absent in the fresh water return rapidly to the area after discharge is cut off.

The general conclusions to be derived from the biological studies are that there are both harmful and good effects of the inflow of fresh water to St. Lucie. There is no doubt that the water becomes brown and turbid, and that certain few species leave the area, but, similarly, there is no doubt that a tongue of salt water exists along the bottom and that most of the fishes in the area involved can withstand very low salinities. In recognition of the temporary loss of fishing for a few species during high flows, Dr. Gunter and Mr. Hall recommend that discharges be held below 3,500 cubic feet a second whenever possible within the necessary operations for flood control.

The biologists are firm in their opinion that a certain amount of fresh water is beneficial to the estuary because it enhances the production of forage fishes upon which the others feed, and also young game fishes. They have made their recommendations based upon the present evidence. The above studies show that flow of fresh water into the St. Lucie is not the unmixed scourge it has been reported to be and that there are appreciable beneficial aspects of such releases of Lake Okeechobee's waters. Since these waters must be released if the lake is to be regulated, as required by the laws of the United States Congress, it is good to know that these actions also have their beneficial aspects.

Appendix B

DEPARTMENT OF THE ARMY
OFFICE OF THE CHIEF OF ENGINEERS
WASHINGTON, D.C. 20314

2 March 1970

Honorable Paul G. Rogers
House of Representatives
Washington, D.C. 20515

Dear Mr. Rogers:

This is in reply to your request of 27 January 1970 for information relating to the ability of the Corps of Engineers to regulate safely the water level of Lake Okeechobee, as discussed in a letter from Mr. M. Heuvelmans dated 12 January 1970.

Mr. Heuvelmans cites changes in the runoff and discharge characteristics of the Kissimmee River resulting from project canalization and indicates a strong concern over the possibility of occurrence of a 30-foot hurricane tide in St. Lucie Canal and its destruction of the city of Stuart, Florida.

Prior to the construction of project works on the Kissimmee lakes and canalization of Kissimmee River, the recorded levels of the major basin lakes fluctuated quite regularly on an annual basis up to 5 or 6 feet and as much as 10 to 12 feet during consecutive periods of alternate floods and droughts. Inadequate outlet capacity for transfer and removal of floodwaters resulted in major floods in the basin in 1947, 1948, 1953, 1959, and 1960 with up to 500,000 acres flooded for durations up to 118 days. The flood damages prevented by a recurrence in the basin of any of these flood conditions would exceed $21 million at today's price levels.

Seasonal regulation of the Kissimmee lakes within a desirable range of 2.5 to 3.0 feet to meet project purposes requires fairly rapid removal of heavy rains and runoff, such as occurred in the September-October 1969 period noted by Mr. Heuvelmans. Discharges from spillway structure S-65E into Lake Okeechobee ranged from about 6,000 c.f.s. in late September up to 20,000 c.f.s. and slightly higher for the period 3-7 October, gradually returning to about 5,000 c.f.s. in early November. The Lake Okeechobee stage rose above schedule on 3 October; regulatory releases were initiated to maximum practicable capacity of Caloosahatchee River (Canal 43) but not to the maximum of St. Lucie Canal in order to avoid sedimentation problems in the Stuart area. The lake reached a stage of 16.6 feet, 1.1 feet above schedule, in early November and was lowered to schedule by early December. In that 30-day period the total outflow from the lake exceeded 730,000 acre-feet.

The possibility that a 30-foot hurricane tide could occur in the St. Lucie Canal is extremely remote. The question of whether hurricane tides generated in Lake Okeechobee could be translated down the canal and to what extent was investigated in detail in 1967 by Dr. Garbis Keulegan. The problem considered a 17.5-foot m.s.l. lake stage with a severe maximum probable hurricane occurrence on the lake. A wind tide elevation of 26 feet m.s.l. would result at Port Mayaca. However, because of the length of St. Lucie Canal, the effect of canal bank and bottom friction, and lateral overflow along the canal, Dr. Keulegan determined that the peak elevation that would be reached at St. Lucie Lock and Dam would be 22.5 feet m.s.l. Some structure overflow would occur for a brief period but without any resultant damage downstream.

Pending appropriation of project funds, the scheduled competition of project works in the north and northeast shore areas of the lake in the next 2 to 3 years will permit raising the regulation range of the lake to the 1954 authorization levels 15.5 to 17.5 feet m.s.l. The design of the levee system and project works around the lake for those higher levels is based on several combinations of flood stages and hurricane intensities. The recurrence of hurricane Laurie in late October 1969, had it crossed the lake, would not have created any problems of the magnitude envisioned by Mr. Heuvelmans. It is because of the memory of the 2,500 dead in the 1928 hurricane that such protective measures were planned, authorized, and built into the project.

In summary, reduction of flooding and the prevention of flood damages in the Kissimmee River Basin requires the capability for rapid removal of flood runoff. With the approaching completion of a short reach of Canal 38 that capability will have been provided. The capability to store more water safely in Lake Okeechobee at higher regulation levels for the rapidly expanding needs of southern Florida will be realized within a few years. As noted above, in the planning and design of project works involving the safety of the general public all necessary precautions are taken by the Corps through detailed studies and testing of various critical conditions and situations that have been known to occur or could possibly occur.

If any additional information is needed please let me know.

Sincerely yours,

1 Incl   **LEONARD EDELSTEIN**
Cy ltr fm Mr. Heuvelmans   Colonel, Corps of Engineers
dtd 12 Jan 70   Assistant Director of Civil Works
                                 for Atlantic Divisions

Appendix C

Honorable John D. Dingell
House of Representatives
Washington, D.C. 20515

Dear Mr. Dingell:

This is in further reply to your recent letter, inclosing a letter from the Martin Anglers Club of Stuart, Florida concerning the muddy appearance of the Kissimmee River.

The discoloration of the waterway about which Mr. Heuvelmans is concerned is due to construction activity related to the Central and Southern Florida Flood Protection Project. The condition is temporary and will clear up in a short time following completion of the work. Operation of the control structures associated with this project, as suggested, will have no effect on the existing temporary muddy condition of the river.

The project design and the construction activities have been closely coordinated with both State and Federal Fish and Wildlife and Water Pollution Control Officials. It is realized by all concerned that during this stage of construction less than ideal conditions must be tolerated in order to accomplish the work that will provide extensive benefits to anglers, boaters and sightseers in this area. The project plan provides for the preservation of natural features conducive to fish propagation and fishing activities and also provides for beautification features to benefit the aesthetically minded boater. All planning and work on this project have been coordinated with other interested agencies in accordance with existing Laws, Regulations, and Executive Orders.

As previously stated the present muddy condition of the waterway is temporary and with the completion of construction will clear up.

Sincerely yours,

WILLIAM F. CASSIDY
Lieutenant General, USA
Chief of Engineers

Appendix D

STATE OF FLORIDA
GAME AND FRESH WATER FISH COMMISSION
P.O. BOX 1840
VERO BEACH, FLORDA 32960

Dr. O. E. Frye, Jr., Director

H. E. Wallace, Assistant Director

August 15, 1969

Mr. Martin Heuvelmans
P. O. Box 130
Stuart, Florida

Dear Mr. Heuvelmans:

In answer to your letter of July 30, 1969, I can provide you with the following information.

I am enclosing a copy of our trip report of June 13, 1969 on the Kissimmee River.

In reference to your questions:

1. The culverts coming from the dam seem to keep the oxbows open until the old river channel crosses the canal. As you

know from there on down to the next structure the oxbows are open at both ends but they don't appear to flush themselves very well. In certain areas heavy silt deposits and aquatic weed jams are occuring.

2. To the best of my knowledge there are no underwater berms on the Kissimmee River. We have probed for them in a number of locations and have been unable to find them. This can also be noted in our trip report of June 13 (attached).

3. The so called fish breeding canals have been built to the specifications we suggested. In field reviews since their construction we have found a number of problems with them. 1. We feel the fish being produced in them will stay in them and not enter the Kissimmee Canal. 2. Some of them have filled completely with hyacinths. 3. Many of them have been coated with silt from the construction going on upstream and these will be low in production.

4. As you may know, we the fish and wildlife agencies would have much preferred a floodway rather than a channel. The coordination of the channel project has been satisfactory, but there is little left to save.

5. Upon completion of the project the fish and wildlife resources of the Kissimmee River Valley will have suffered greatly. Attached is a speech that I gave to the Florida Board of Conservation Water Resources Department in which I explained in more detail what we feel is happening.

I personally feel, possibly optimistically, that in the future the Kissimmee River floodplain will be repurchased by the state to restore the nutrient trapping capability of the floodplains. If this is not done, Lake Okeechobee will become Florida's second Lake Apopka.

Sincerely yours,

FLORIDA GAME AND FRESH WATER
FISH COMMISSION
Larry R. Shanks
Project Leader

LRS/mr
Enclosures

Appendix E

Nov. 3, 1958

Mr. Robert T. Bair,
St. Lucie-Indian Rivers Restoration League
Jensen Beach, Fla.

Reference: Dist. Eng. (Jacksonville) letter dated October, 1958. Re: proposed revision in Okeechobee flood control project.

Dear Mr. Bair:

I have delayed replying to your request for comment on reference letter, pending consultation with other committee members and our technical advisory group.

The very general and non-specific terms of the letter make it subject to various interpretations as to just what is proposed and what may be eventual results of the revisions and new construction outlined. To a casual first reading it appears to hold promise of substantial relief from our present estuary troubles, but on more careful study of its underlying implications it presents a quite different picture, as we see it.

The most significant feature of the letter, although it avoids saying so in so many words, is the unmistakable implication that in official view of the Jacksonville District office, a Floodway (Third Outlet) or other alternate means for emergency lake discharge to the lower west coast does not merit further considerations and should be deferred indefinitely. This comes as a surprise because this view is widely at variance with my distinct understanding of the sense of our Jacksonville Conference to which the letter refers, is contrary to the sense of the subsequent formal statement of Flood Control District governing board as quoted in the public press shortly thereafter.

I refer specifically to Mr. Eden's presentation of preliminary studies outlining alternative routes for additional lake outlets southward to the lower west coast to serve the dual purpose of (1) a safety valve floodway for the rapid and comparatively harmless escape of emergency flood waters to coastal areas which nature designed to receive them, (2) to facilitate the controlled supply of dry season needs to large areas not now served (e.g. Everglades National Park and the Devil's Garden and Immokalee areas, etc.).

Instead, it is apparently now proposed to continue indefinitely to rely for lake regulation upon the present expedient of wasteful (and damaging) discharges into highly developed coastal areas via the St. Lucie and Caloosahatchee Canals, but with the palliative measures in the form of enlargement of Caloosahatchee with implied intent of diverting to Caloosahatchee a part of those waters now channeled to St. Lucie Outlets.

This is an expedient which at best can be only partial and can afford only temporary relief from some of the presently highly localized pains, and which would defer indefinitely a positive approach to the one measure (a major southward floodway) that can contribute most to an effective permanent cure for a wide variety of ills now prevalent in all areas of the Flood Control District.

Considerable enlargement of the Caloosahatchee is doubtlessly a local necessity in Hendry and Lee Counties because Caloosahatchee presents discharge capacity not large enough to take care of flood waters originating in its own tributary watershed, irrespective of lake discharge. For that reason, Hendry and Lee Counties interests will certainly benefit by Caloosahatchee enlargement up to the discharge capacity required to handle the local run off.

How far beyond that point it is now proposed to carry the Caloosahatchee enlargement is not stated but the context clearly implies an intent to make it at least large enough to take lake discharges (in addition to local runoff) approximately those now channeled to St. Lucie or perhaps greater and make the Caloosahatchee the primary regulatory channel instead of the St. Lucie. Such a diversion would, if effected, afford a limited measure of relief to the St. Lucie estuary, but it may be confidently predicted that if carried to the extent necessary to solve the regulatory problem, it would in due course create a condition in the Ft. Myers estuary similar to that now current in the St. Lucie estuary and incite justifiable protest from Ft. Myers no less vehement than those now current from Stuart. We note especially the statement that:

"The revised plan would also mean that when discharges were required by way of St. Lucie, it would be made insofar as is possible at a comparatively low rate about 3500 to 4000 cfs. This would eliminate the canal erosion and sandbar deposition in the estuary".

Any reduction in lake discharges, however small, would of course help in some measure, but even if it proved possible to limit St. Lucie discharges to 4000 cfs (which is not assured) this would certainly not eliminate the erosion and deposit problems because:

1. St. Lucie erosion is now influenced to a greater extent by the undercutting action on vertical banks of stern waves of passing water craft, plus the caving action from heavy rainfall, that it is influenced by the scouring action of high current velocities, and-

2. A very considerable part of our St. Lucie estuary siltation is not sand, but is rather soft organic or colloidal matter that does not settle out in the comparatively still waters of Lake Okeechobee nor in the canal but remains in solution or suspension until precipitated by contact with tide water in the estuary.

The foregoing comments have been directed principally at the one specific local problem of our St. Lucie estuary damages. If that were the only problem involved it might appear expedient to accept the palliative measures proposed, as advantageous to our local interests, because they would obviously afford some measure of relief for our present estuary pains, and in some measure benefit the economic interest of Martin County as a whole.

Martin County has other problems of equal, though perhaps less obvious, importance, all of which are related to the major problem of district wide, even statewide concern, namely the conservation of water resources now wasted, and the needful distribution to areas not now served. The key factor to the solution of that problem, to paraphrase a recently published FCD statement, is the efficient management, control and regulation of Lake Okeechobee and related conservation areas. That in turn involves the inescapable necessity for a major floodway or other alternative "Safety Valve Outlets" to the southward of the lake discharging into the Gulf in the general vicinity of Everglades National Park.

That view, as you know, has been repeatedly expressed by various individuals from time to time for many years, but rejected from serious consideration by responsible state and federal authorities on the ground that it was economically unfeasible. This was a justifiable decision in the economic conditions which existed some 20 years ago (1938) when the present project plans were conceived. It was justifiable perhaps 10 years ago when the present project plans were formally authorized.

The economic criteria of ten years ago have been made completely inapplicable to conditions of today by the explosive developments that have taken place during the past decade. Many projects then considered economically infeasible, have now become inescapable economic necessity. Among these, the southern floodway merits highest rank because of the beneficial influence it can exert in the solution of all problems related to the mounting needs for fresh water conservation and dry season distribution as well and the mounting demand for flood protection in a rapidly expanding population and economy.

It may be superfluous, but it is pertinent to review here certain obvious fundamentals upon which this view is based and which we believe must govern any program aimed at truly effective control and distribution of Lake Okeechobee water.

1. Safety to human life is paramount. Hurricane breaking of levees must be prevented at all costs.

2. Present safe storage capacity in Lake Okeechobee is insufficient to meet current needs. It must be increased.

3. Progressive increases in safe lake storage capacity must be accompanied by commensurate increases in emergency outlet capacity. Conversely, increases in available outlet capacity permit commensurate increases in safe storage capacity.

4. With any given height of containing levees, the larger the escape outlets the greater the effective safe storage capacity and vice versa.

5. Discharge now available is inadequate for efficient lake regulation.

These are obvious generalizations whose truth is, we believe, beyond argument. That they are recognized by the Jacksonville District Engineers Office is clearly indicated in the reference letter. Otherwise there would be no logic in the proposed linking of Caloosahatchee enlargement with increased levee heights in the proposed first stage increment in lake storage capacity, and no logic in the statement regarding further possible storage increment that ". . . the second stage would require an appreciable increase in outlet capacity."

With the foregoing fundamentals in mind, and without disputing the essential desirability of the proposed levee additions, it is our considered belief that the safe storage capacity of the lake could now be significantly increased without increase in existing levee height if adequate safety valve outlets were now available. This view stems from consideration of the following controlling factors which are generally accepted by all informed agencies including Jacksonville District.

1. Danger of overtopping levees by wind driven hurricane waves of 9 feet or more, dictates an irreducible wave margin which must be maintained between levee crest at its lowest point and the highest mean water level that may be reached during and after a severe hurricane.

2. Hurricane rainfalls up to 15 inches in 24 hours have been recorded over the lake and its tributary watersheds. Result: an immediate rise of more than a foot during the storm. This is more water that the St. Lucie and Caloosahatchee together can now discharge in three weeks. This immediate rise will be followed by a continued rapid rise up to three feet or more within a few days, caused by run off from tributary watersheds.

3. In flood conditions the total discharge capacity of Caloosahatchee and St. Lucie is not immediately available for lake discharge because the lands bordering the canals themselves cause a preemptive flood crest between the lake and the sea which serves as a virtual block against outflow from the lake for a considerable period of time, at just the time when lake discharges are most needed.

Thus we face a situation where capacity of existing outlets available for lake discharge in hurricane conditions is so small by comparison with the volume of water that must be handled that their effectiveness as an instrument of lake regulation approaches the vanishing point at just the time it is most needed.

In the absence of outlets of any significant emergency value, the only safe recourse is the present practice of holding the lake at so low a level during the hurricane season (the season of greatest rainfall) that the lake can safely absorb the entire volume of water that might be imposed if a major hurricane does occur. This safety valve must be maintained until after the hurricane season is past. It is so low that we enter the ensuing season of least rainfall with a storage reserve which is too small to take care of usage supply needs in the event that replenishment rainfall fails (as it occasionally does fail) to occur in time to avert an acute shortage. It was only the unprecedented mid-winter rain of last winter that averted such an acute shortage within the past year.

If on the other hand, emergency outlets were now provided of large enough capacity to effect a significant reduction in lake levels during the alert period, now afforded by our hurricane warning services, the normal summer safety level for existing levees could be correspondingly raised and the threat of acute shortages correspondingly reduced. The larger the safety valve capacity the higher the permissible summer safety level capacity and the greater the normal safe storage capacity. This is a generality which holds true not only for the present levees but for any levee conformation which may be provided in the future.

For the foregoing reasons the conclusion is inescapable that additional lake outlets to the lower west coast are a necessity of the present and that they will become increasingly urgent with the current growth in population and in rural, urban and industrial activities and the related mounting demands for conservation and broader distribution of our fresh water resources.

It is inconceivable to us that any other conclusion can reasonably derive from thoughtful consideration of the established facts and controlling factors outlined above, which factors we believe are well known by the District Engineers, and are beyond rational dispute. We find it difficult to believe that the attitude clearly implied in the letter accurately reflects the conviction of the office of origin. We note especially the very iffy phrasing of specific references to the floodway and the obvious avoidance of any specific expression of opinion that a floodway is not now necessary.

It is the firm conviction of this committee that a floodway to the south is a real necessity of the present, and that deferment of whatever action may be required to put it into physical being at the earliest practical data is prejudicial to the public interest.

> Very truly yours,
> Bruce G. Leighton, Chairman
> Martin County Water
>   Conservation Committee

Appendix F

CHANNELIZATION
(Text of testimony presented by Nathaniel P. Reed,
Assistant Secretary of the Interior)

Mr. Chairman, thank you for providing me with the opportunity to appear before this Subcommittee to discuss the scope and effects of stream channelization within the United States. In reading your request for us to appear it seems the Subcommittee is searching for answers to three basic questions.

1. Does our agency have comments and data which demonstrates some of the adverse environmental effects caused by stream channelization?

2. It is possible to protect and enhance our environment while still providing needed flood protection?

3. Should some sort of moratorium be placed on the funding and construction of stream channelization projects until such time as new procedures and policies are established?

First, the problem:

I was stunned last week when a copy of the Corps of Engineers' booklet "Water Resources Development by the U.S. Army Corps of Engineers in Delaware" dated January, 1971, crossed my desk showing completed public works projects for that state. This report indicated there is only one major stream in Delaware that has not been channelized at least in part—the Appoquinimink River near Odessa—and the Corps has authorization to work on that.

In the recent controversy concerning the channelization of the Alcovy River in Georgia, it was learned that the Soil Conservation Service under P.L. 566 has plans for alteration of nearly every watershed in Georgia. Reviewing the status of small watershed projects in the southeastern states alone, we found that as of August 1, 1969, 1,119 applications for watershed assistance have been received covering 122,620 square miles. Of that number, 638 have been authorized for planning and 428 have been approved for installation. Estimates indicate that projects in just this one program will involve the alteration of over 25,000 miles of stream channels to obtain flood protection and drainage objectives. These alterations will adversely affect from 25,000 to 60,000 acres of stream habitat. A conservative estimate of the wooded wildlife habitat damaged or destroyed by these alterations would be about 120,000 acres and could exceed 300,000 acres.

After an inquiry with the field staff in preparation for this statement, we found these trends occurring throughout the nation, and I should indicate that the Corps of Engineers and the Soil Conservation Service agencies are not the only agencies engaged in these practices. I am also aware of criticism of some facets of Bureau of Reclamation projects as well.

Second, the question of adverse environmental effects:

Stream channelization projects usually entail changing the physical shape of the stream bed and bank, regulating natural stream flow patterns, and impounding or modifying the flood plain. If the emphasis on these practices continues, the ultimate result will be the destruction or serious degradation of valuable and irreplacable natural resources, including stream fisheries and wildlife in many bottom lands and water courses.

Stream channel alteration under the banner of channel "improvement" for navigation, flood reduction, and agricultural drainage is undoubtedly one of the more, if not the most, destructive water development or management practice from the viewpoint of renewable natural resources. These alterations are carried out in varying degree, with a corresponding variation in damages to stream ecology.

Stream channel excavation which increases the width and depth and changes alignment of a natural channel is the most damaging of these practices. Following in descending order of their detrimental effects are extensive clearing and snagging with dip out, clearing and snagging, minor snagging, and selectively cleared stream channels and or floodways.

Channelization or other stream alteration practices destroy the balance of space and associated life supporting elements. The effects of stream alteration on fish and wildlife is somewhat analogous to the impact of Hurricane Camille on the human population along the Gulf Coast.

After the hurricane (or after stream alteration) the space still remains; however, the elements within the space which support vigorous and thriving populations are no longer immediately available or arranged in a fashion so as to be usable. Fortunately, man has the capability and desire to rebuild his environment following such disaster. Fish and wildlife lack this rebuilding potential; therefore, the organisms must evacuate the damaged or destroyed habitat or perish.

I would like to submit for the written record the following series of photographs depicting stream channelization activities and the accompanying field reports describing the photographs and interpreting them. In addition, I invite your attention to several slides that will be shown depicting one or more stream-channelization projects in my home state of Florida.

Studies conducted by the North Carolina Wildlife Resources Commission evaluated the effects of channelization on fish populations in eastern North Carolina streams. These studies showed that the production of game fish species was reduced by 90 percent following channelization. They further demonstrated that this loss is a permanent loss because normal maintenance procedures preclude the possibility of recovery of the stream's normal productivity.

A similar but unpublished sample relating to the fish population before and after channelization in Tippah River was obtained by the Mississippi Game and Fish Commission. Before channelization, a population sample was taken which revealed a total standing crop of 877 fish per acre weighing 241 pounds. Another sample obtained following channelization disclosed a total standing crop of 1,498 fish per acre weighing only 5 pounds. These comparative data shows a 98% reduction in the weight of fish per acre with a 60% increase in the number of fish per acre. The marked increase in the number of fish may be misleading since 99% of these fish were minnows, shiners and darters with a combined weight of 4.4 pounds. Damage to fish habitat brought about by man's alteration of stream channels occurs across the United States. We have studies in Montana, Florida, Missouri and other areas further documenting losses of 80 to 90 percent of stream productivity.

These studies provide shocking and irrefutable evidence of the severe damages to fish habitat and populations in the immediate area of channel alterations.

Additional stream habitat degradation also occurs for some distance downstream from the altered areas. Siltation and turbidity associated with upstream channel alteration and disruption reduces light penetration in downstream waters, particularly during construction and until some reasonable degree of channel stability is achieved. This reduction in light penetration results in reduced photosynthetic activity by aquatic plants which are important links in the food chain. These plants provide a certain amount of dissolved oxygen which is essential to a healthy aquatic environment. As the suspended particles settle out, they blanket large areas of productive habitat, thus seriously reducing or completely destroying the areas capability to provide the essential elements for fish survival and reproduction. To me, this phenomenon is the aquatic version of the dust bowl disaster.

Some channels are constructed for navigation; however, the stated purpose of most channelization proposals is to increase the volume and velocity of flow for flood protection and or drainage. In essence, this is water disposal and not water conservation, which in turn creates instant drought in the channeled area and instant floods in downstream segments. These modifications can create problems in downstream segments which generate the need for more channel alterations. For example, the Corps of Engineers has been authorized to restudy the Tensas-Cocodrie area of Louisiana to determine the need for more extensive flood control and drainage improvements. This study will consider alteration in the entire Bayou Cocodrie, Cross, Black and Buckner Bayou channels which provide the main outlets for two proposed Public Law 566 channelization projects in Concordia Parish Louisiana. Each of these projects depends on the other for full realization of projected joint flood control and drainage benefits, and both will destroy fish and wildlife habitat.

The increase in quantity and speed of flow causes waters to carry a much higher silt load into downstream reaches. Under natural conditions high waters spread out over the seasonally flooded bottom lands and swamps, thus greatly reducing the flow velocity, permitting the settling out of much of the silt load and reducing turbidity. These overflow bottom lands and swamps, which are highly productive of timber and wildlife, are nature's own floodwater retarding structures. They may also perform other functions, such as recharging ground water storage areas, filtering and purifying surface flows, and controlling eutrophication of downstream waters by removing and utilizing nutrients.

The specific impact of channel alterations on the quantity and quality of bottom-land wildlife and waterfowl populations has not been the subject of intensive study. However, it is clearly evident to anyone who understands the rudiments of biology that habitat disruption and destruction of the magnitude caused by stream channel alterations result in serious losses to waterfowl and other bottom-land wildlife.

Stream channelization results in a direct loss of woodland habitat through right-of-way clearing for equipment access and spoil disposal. Some mitigating of this loss occurs when wildlife plantings are placed on the modified areas.

Channel alteration accelerates the removal of surface waters from swamps and marshes and greatly reduces the frequency and duration of seasonal flooding of other wooded bottom lands. Seasonal and permanent surface water, which are essential factors in maintaining these ecological units, are greatly decreased or eliminated. Loss of this surface water will allow encroachment of undesirable underbrush, inhibit growth and reproduction of desirable vegetation, reduce aquatic and wetland habitat, eliminate swamp refuge or escape areas, and significantly reduce or eliminate waterfowl utilization. Our experience indicates that installation of flood control and drainage channels encourage the acceleration and construction of smaller private drainage projects that further reduce the quantity and quality of wooded bottom-land wildlife habitat.

I think we are kidding ourselves if we do not admit that the vast majority of stream channelization has had a devastating effect upon our nation's waterways. We could spend all day detailing the endless miles of streams slated for additional modification by one agency or another. But that will not solve an admittedly serious problem. What is needed is a complete rethink and redirection by the men who are designing and constructing the projects.

This leads us into the second question . . . is it possible to protect and enhance our environment while still providing needed flood protection?

While the demand increases for wild and scenic rivers, for fishing, hunting, swimming and open space, and environmental quality, our supply is rapidly decreasing. The philosophy to date has been that as people move into and develop the river flood plains they demand flood protection, water for domestic and agricultural uses, and navigation to import and export the goods of our consumer-oriented economy, and have sacrificed our rivers and streams to accommodate these apparent demands.

Even though we spend millions of dollars each year for ditching, dams and diking of our rivers and streams, the flood damage throughout the nation continues to rise. Perhaps our philosophy has been misdirected. We have some federal agencies charged with doing a job which involves environmental destruction and others charged to protect the environment in continuous conflict. A redirection would involve a land use philosophy which by necessity would include flood plain delineation. After the flood plain has been defined, then flood plain zoning practices must be implemented which allow land use compatible with periodic flood cycles: such land uses in the flood plain would involve fish and wildlife production, open space pastures, parking lots, recreation areas, and other demands for space which can withstand temporary flooding. This redirection of land use practices would not only aid in saving fish and wildlife environmental quality, but would also reduce insurance losses and other losses during flood periods. We realize this will not eliminate the damage, but it would reduce the economic losses to our society.

The Department of the Interior definitely feels that there are ways that the environmental quality of the nation can be protected while still providing needed flood protection. The flood suggestions would aid in this endeavor:

1. Allow land owners to reduce their taxable acreage by the amount of land they have in wetland areas as long as it remains in its natural condition. This could include flood plain hydric hammocks and marshes. The fish and wildlife resource values of these areas must be approved by a state or federal environmental agency prior to their acceptance. Furthermore, a land-owner commitment that these lands will remain in their natural condition for at least a 10-year period of time would be necessary.

2. Encourage Congress to pass legislation establishing a green belt of vegetation which must be left along rivers and streams to protect the river ecosystem from erosion, as well as sustaining fish, wildlife, and environmental quality.

3. Zone flood plains so that whatever use is made of the land it should be able to withstand temporary flooding. There are certain land uses which can serve our society and still be compatible with occasional flooding.

4. A complete revision of Public Law 566 to incorporate purchase of lands for fish, wildlife, public access, recreation, environmental quality, and other needs of our modern society.

These recommendations alone, however, will not suffice. Existing uses and commitments in the flood plain zones necessitate some continued project works.

The National Environmental Policy Act of 1970 was a meaningful step toward weeding out the truly environmental destructive proposals. It has however, one serious flaw. The act is basically reflective in nature and not designed to function as an effective early warning system for society's decision makers. Project review is not accomplished until such time as the proposed project design has been for all practical purposes, decided upon. Our experience to date has been that it is extremely difficult to effect project revision when the project has arrived at the Council on Environmental Quality for final review.

Proper input into the project design from its inception by qualified, knowledgeable professionals in the environmental field is essential.

Under existing procedures, the Bureau of Sport Fisheries and Wildlife functions only in an advisory capacity to the other agencies authorized to design and construct stream channelization projects. These agencies are under no cumpulsion to integrate our recommendations into the project design nor are our objections overriding under existing procedures.

What is clearly called for is a reallocation of agency priorities. The Department of the Interior and the Environmental Protection Agency should be given a much stronger and more meaningful voice in the development of project design. President Nixon's proposed super-agency, the Department of Natural Resources, would accomplish some of this goal. However, additional enabling legislation may also be needed. The Department of the Interior, working with EPA, has the expertise to insure that all stream channelization projects are designed and constructed in a manner to minimize their destructive impact upon the natural resources. It is time that the Congress gave the environmental agencies the leadership role in determining project design. Make us a leader rather than a frustrated follower. A large portion of the morale problem within my department is the result of rarely being listened to when we offer relevent recommendations to other agencies on this problem. It is discouraging for our biologists and field personnel to stand by helplessly and watch the wetlands resource succumb to the dredge bit or dragline bucket with little or no regard for the natural system.

In addition, something must be done and done soon about our manpower problem.

The Bureau of Sport Fisheries and Wildlife has become greatly overcommitted in recent years under its River Basin Studies programs. The workload has been increasing at a rapid rate and the Bureau has not been able to obtain the funds for manpower necessary to keep pace.

And now to the third question I posed in my opening remarks. "Should some sort of moratorium be placed on stream channelization activities at the present time?

In answering this question I must first tell you quite frankly that it has been the observation of the majority of our personnel that those agencies engaged in stream channelization activities are still largely paying nothing more than lip service to earnest environmental protection. We have yet to detect any substantive departure from the practices of yesteryear by these agencies, and I believe the record will clearly support these conclusions. This does not apply to the Bureau of Reclamation who has been operating under Departmental directive for 5 or 6 years. In view of our continuing problems in this vital area, it is my belief that the following items should be given careful consideration as a means to further protect these rapidly vanishing wetland systems:

1. A complete review of all river and stream channelization projects should be initiated by the Council on Environmental Quality working in cooperation with the Department of Interior and the Environmental Protection Agency. This review should be directed to the possible need for project redesign or project deauthorization. If the supporting agencies fail to take this review seriously, and if nothing more than lip service is paid to redesigning these projects, then I would welcome the opportunity to reappear before this committee to discuss the imposition of a complete moratorium on all such projects until these reviews and necessary project revisions have been completed.

2. The Congress should consider authorizing the Council on Environmental Quality to recommend deauthorization of all projects, including presently authorized projects not started within five years of Congressional authorization.

3. Policy guidelines should be developed and enforced for compatible uses of flood plain ecosystems. Acceptable land use practices for these systems should be determined by the Department of the Interior working in conjunction with other agencies. The environmental agencies however, should have final veto power over project designs determined by the Department to be too costly in environmental currency.

Mr. Chairman, that concludes my remarks to the Subcommittee. In closing I would like to submit for the written record the report prepared by Mr. E. C. Martin of the Department and quoted at length in my remarks on the adverse effects of stream channelization on our wetlands resources. Mr. Martin is to be commended for an excellent paper.

I apologize for the length of my remarks. Stream channelization is a matter of deep concern to me personally, and I wanted to give it the discussion I felt was needed.

    ✿   ✿   ✿   ✿   ✿   ✿   ✿   ✿   ✿   ✿

Appendix G

IDAHO FISH AND GAME DEPARTMENT

Statement of the Idaho Fish and Game Department presented at Corps of Engineers Public Meeting on Lower Clearwater River at Lewiston, Idaho November 19, 1970.

The Idaho Fish and Game Commission by resolution adopted July 29, 1969, has officially opposed any further dam construction in the Clearwater River drainage. A copy of this resolution is attached for inclusion in the record of the present meeting.

Importance of the Clearwater River drainage to the fish and wildlife resources of the State of Idaho and the Pacific Northwest cannot be overemphasized. The drainage contains the largest elk herd in the state and in the nation. Idaho produces an estimated 55% of the total Columbia River summer steelhead run. Approximately half of these fish are produced in the Clearwater River System. Spring and summer Chinook salmon have recently been successfully reintroduced into the drainage. A potential exists for the establishment of salmon runs which could rival the Salmon River Chinook salmon sport fishery in popularity and economic

213

value to the state. A significant smallmouth bass fishery also exists in the lower reaches of the Clearwater River.

A major portion of these invaluable existing and potential resources have already been eliminated or seriously jeopardized by construction of Dworshak Dam. Approximately 50% of the anadromous fish production area in the entire Clearwater River drainage has been eliminated. The North Fork Clearwater River drainage has been completely removed from the Idaho salmon and steelhead fishery. The extremely popular "resident" fishery in the North Fork which was maintained by juvenile steelhead will be nonexistent after 1971, when progeny of the last of the steelhead runs moving above the dam leave the drainage. It will be many years before a comparable resident fishery can be reestablished above Dworshak Reservoir, if at all. Reestablishment of Chinook runs in the North Fork Clearwater river as has been accomplished in other portions of the Clearwater River system is precluded by construction of Dworshak Dam. Dworshak Reservoir will inundate about 15,000 acres of critical elk and deer winter range whose greatest value is its ability to keep animals alive during emergency periods.

Mitigating measures in the case of big game are inadequate. Out of a total of 50,000 acres of adjacent land needed for instensive range management to maintain existing populations there is now only the possibility of obtaining 5,000 acres. Severe big game losses furing heavy winters will almost certainly occur after impoundment of Dworshak Reservoir.

An attempt will be made to replace the steelhead run by artificial production below the dam. Success of this venture is by no means assured. Regardless of hatchery success, it is impossible to replace the lost production and fishing area.

As previously mentioned, reestablishment of the resident fishery above the reservoir will take many years and may not be possible if undesirable fish species during the interim occupy the niche left vacant by juvenile steelhead. No effort has been made to replace the potential Chinook production which has been lost due to dam construction.

The losses which have already occurred, and others that will occur after impoundment, were predicted by the concerned fish and wildlife agencies prior to authorization of Dworshak Dam. The Corps of Engineers at that time attempted to minimize the reports of the fish and wildlife agencies and continued to maintain that

Dworshak Dam would cause infinitesimal damage to the fish and wildlife resources.

Dworshak Dam is now an accomplished fact and most of the really tragic losses to the fish and wildlife resources that have been and will continue to be experienced are irreversible.

The Idaho Fish and Game Department feels very strongly that the adverse effects of the originally authorized project should not be further compounded by supplemental construction or project operation that would be detrimental to fish and wildlife resources or public enjoyment of these resources.

A regulating dam below Dworshak Dam will not be necessarily reducing of river level fluctuations due to peaking operations. The reregulating dam can itself be operated for peaking power and would produce river level fluctuations up to the maximum that public opinion will tolerate—which will be exactly the case if Dworshak Dam is operated without regulation.

Passage over and through additional dams and reservoirs in the Clearwater River would inflict further mortalities on upstream and downstream migrating anadromous fish runs already in critical condition as a result of passage through Corps projects on the Columbia and lower Snake Rivers. There is every indication that these runs would not survive if they were subjected to additional stresses.

Construction of a reregulating dam or dams would create more barriers to migrating fish with inevitable resulting mortalities as fish pass over or through the structures. Reservoirs with attendant predator problems would be formed through which all Clearwater River downstream migrants would have to pass. Steelhead smolts from Dworshak Hatchery would be released directly into the reservoir. Smallmouth bass spawning would be reduced or eliminated by fluctuating reservoir levels. There would be interference with collection of Dworshak Hatchery adults by full-peaking releases and disruption of wild fish passage in the main Clearwater would be very likely. A potential for nitrogen supersaturation would always be present with the creation of slack water immediately below Dworshak Dam.

It has not been possible to maintain meaningful salmon and steelhead fisheries in Columbia Basin impoundments. Experience has also shown that Snake River smallmouth bass fisheries have been seriously reduced upon impoundment of the river. When construction of Lower Granite Dam in the Snake River and Dworshak Dam in the North Fork Clearwater River is completed, the main

Clearwater River will be the only remaining reach of water where the Clearwater River steelhead and salmon runs can be fully utilized by Idaho fishermen. Impoundment of a major portion, or conceivably all of this river section as now under consideration, would not only create adverse effects on existing and potential native fish populations themselves, but seriously limit or reduce to insignificance the fisheries on these populations. The potential contribution to Idaho of the recently constructed multimillion dollar Dworshak Hatchery and the Kooskia National Fish Hatchery would be largely negated by lack of suitable fishing water necessary to fully harvest production from the hatcheries.

A fluctuating reservoir would eliminate vital summer cover for upland bird populations now common in the area and result in a decrease of these populations. There will obviously be no need for additional slack water recreation opportunity in the area after impoundment of Dworshak and Lower Granite reservoirs. Present recreational use on the free flowing river would be drastically reduced by a fluctuating reservoir.

There is little doubt that operation of Dworshak Dam at full peaking capacity without reregulation would have serious adverse effects on aquatic habitat, fish populations, recreation, and other downstream river uses. In the existing authorization, however, there is no requirement or provision for operating the project at full peaking capacity. If the original justification was accurate, the project can be operated on a reduced peaking schedule and still be economical. It is the position of the Idaho Fish and Game Department that peaking operations should be carried out only to the extent that they do not interfere with other uses of the river. The allowable magnitude of these operations can only be determined by experimentation and detailed studies after the project is in operation.

Our department from the beginning has been in close contact with the Corps of Engineers concerning the need for the design and operation of variable level outlets in Dworshak for water quality and temperature control. As previously expressed to the Corps, the Department's primary concern is that existing fish populations do not incur still additional project associated losses as a result of reservoir water releases. To insure these losses do not occur, first priority should be given to demonstrating that the theoretical capability of duplicating existing water temperatures and quality can be achieved under actual conditions. Further manipulation for possible enhancement should be undertaken cautiously and only if indicated after full assessment of all possible effects on all fish populations that will be affected.

The history of water development is replete with examples of a piecemeal approach to the destruction of rivers in the cause of full, so called, economic development. Dworshak Dam, and now the proposal for an additional dam to fully utilize Dworshak Dam, are first steps in the replay of a familiar pattern. Installation of the presently contemplated six generating units and a reregulating dam to accomodate peaking flows from these units does not represent utilization of full peaking capacity at Dworshak Dam. Installation of a large number of even additional units is economically feasible. Full utilization of this further capacity would again require additional reregulating dams. From complete impoundment of the lower Clearwater River by reregulation dams it is only a series of short steps to additional tributary storage and eventual complete impoundment of the drainage when remaining free flowing stream resources have been reduced to insignificance.

Our Department is convinced that the present issue is not merely whether there should or should not be regulation for Dworshak Dam. The issue is actually whether the Clearwater River drainage should be fully developed for hydropower or maintained in a free flowing condition for recreational use and public enjoyment of river associated fish and wildlife resources. The Department is of the firm conviction that the public interest would best be served by retaining free flowing portions of the Clearwater River drainage in their existing condition. Alternative sources of peaking power are available in some cases admittedly at an increased cost. There are, however, no alternative sources that can replace at any cost a free-flowing river with its associated fish, wildlife, and recreational resources, and the social benefit resulting from these resources.

For the above reasons, the Idaho Fish and Game Department is unalterably opposed to any further dam construction in the Clearwater River drainage. We are also opposed to any operation of Dworshak Dam that will further compound losses to the fish and wildlife resources which have been and will continue to be experienced as a result of the construction of Dworshak Dam.

John R. Woodworth, Director
Idaho Fish and Game Department

References

Comprehensive Report, 1947—on South Central Florida Flood Control District. District Office, U.S. Army Corps of Engineers, Jacksonville, Fla.

Partial Definitive Project Report, Central and South Florida Flood Control and other projects. Serial 19, March 28, 1955. District Office, U.S. Army Corps of Engineers, Jacksonville, Fla.

Fresh Water in St. Lucie Estuary—Good or Bad? Serial 53, May 12, 1960. District Office, U.S. Army Corps of Engineers, Jacksonville, Fla.

Study of St. Lucie Estuary. Interim Report (C23 & C24), October, 1958, edited by Robert M. Ingle, Dir. of Research, Florida Conservation Board and submitted to South Central Flood Control District, West Palm Beach, Fla.

Survey of effects of release of water from Lake Okeechobee through the St. Lucie Canal. Interim Report, January, 1954 of James F. Murdock, Marine Laboratories, University of Miami, Coral Gables, Fla. for the U.S. Army Corps of Engineers.

Biological Investigations of the St. Lucie Estuary (Florida) in connection with Lake Okeechobee discharges through St. Lucie Canal. A report to the U.S. Army Corps of Engineers, 1959, by Gordon Gunter, Marine Laboratories, Ocean Springs, Miss.

Martin County, Florida Damages to Coastal Communities from South Central Florida Flood Control District Operations. A report (early sixties) prepared by the Martin County Water Conservation Com-

218

mittee at the request of the Martin County Board of Commissioners, Stuart, Fla.

Detailed Report on Fish and Wildlife Resources in relation to the Corps of Engineers Plan for Development of Kissimmee River Basin, Florida. May, 1957. Office of River Basin Studies, Fish and Wildlife Service, U.S. Department of the Interior, Vero Beach, Florida.

Report on Winterfowl Population of Kissimmee River Valley in relation to Hydrology, Topography, Distribution of Vegetation and proposed Hydrological Regulations. July, 1957. Florida Game and Fresh Water Fish Commission, Tallahassee, Florida.

Channelization of the Kissimmee River. A report, December, 1968, by Jon Buntz, Regional Fisheries Biologist, Game and Fresh Water Fish Commission, Tallahassee, Florida.

A Synoptic Survey of Limnological Characteristics of Big Cypress Swamp, Florida. May, 1970. Prepared by John A. Little, Robert F. Schnieder and Bobb J. Carrol, Federal Water Quality Administration, U.S. Department of the Interior, Atlanta, Ga.

Report of the 5 man Special Study Team (Loveless Report) on the Florida Everglades. August, 1970. A report by Charles M. Lovless, Chairman and Assistant Director, Denver Wildlife Research Center, U.S. Bureau of Sport Fisheries and Wildlife, Denver, Colorado.

Final Report on Hydrological Reconnaissance of Conservation Areas #1, 2 and 3. January 17, 1971. A report prepared by J. P. Heaney, PhD and W. C. Huber, PhD, of the Department of Environmental Engineering, University of Florida, Gainesville, Florida for the Central and Southern Florida Flood Control District.

Environmental Impact of Cross Florida Barge Canal on the Oklawaha Regional Ecosystem. March, 1970 and January, 1971 statements of Florida Defenders of the Environment, Inc., Gainesville, Fla.

Statement to Governor of State of Florida by Plenary Committee on Water Management in Session, September, 1971. Miami Beach, Fla.

The Kissimmee-Okeechobee Basin. A report, December 12, 1972 to the Florida Cabinet by the Director, Division of Applied Ecology Center for Urban and Regional Studies, University of Miami, Coral Gables, Fla.

Minutes of the Meeting of the Florida Cabinet, December 12, 1972. A transcript, Office of the Governor, Tallahassee, Fla.

Environmental Statement, Central and South Florida Projects (Kissimmee), a summary draft, 5 May, 1971. District Office, U.S. Army Corps of Engineers, Jacksonville, Fla.

Summary of Corps Plans for the Apalachicola by Mike Toner, *Miami Herald,* July 16, 1973.

Environmental Defense Fund, Inc. vs. Corps of Engineers and Generals Resor and Clarke, a transcript of court decision on the Gillham Dam in Arkansas, filed February 19, 1971 (No. LR-70-C-203), Eastern District of Arkansas, Western Division.

Bristow, B. *Report on Land and Water Investigations, Project FW 16-R-7 and R8, 1968 and 1969. Wildlife Views,* Arizona Game and Fish Department, Phoenix, Arizona.

Report on the proposed Ben Franklin Dam on the Columbia River. April 2, 1969. Comments of Fish and Wildlife Service, Regional Office, U.S. Department of Interior to District Engineer, U.S. Army Corps of Engineers, Seattle, Washington.

The Effects of Stream Alterations in Idaho. April, 1969. An analysis by Richard A. Irizarry, Fish Biologist, Idaho Fish and Game Department, Boise, Idaho.

Summaries of studies of measures taken by the Corps of Engineers to reduce losses of salmon and steelhead in Columbia and Snake Rivers. September, 1971. Prepared by National Marine Fisheries Service, U.S. Department of Commerce, Seattle, Washington.

Effect of Little Goose and Lower Monumental Dams on the success of outmigrations of juvenile Chinook salmon from the Snake River, 1970-72. January, 1973. A summary prepared by Howard Raymond, National Marine Fisheries, U.S. Department of Commerce.

Re the Dworshak Dam. November, 1970. A statement by the Idaho Fish and Game Department at a public meeting conducted by the U.S. Army Corps of Engineers at Lewiston, Idaho.

Supersaturation of Nitrogen in the Columbia and Snake Rivers. March 23, 1971. A statement by John A. Biggs, Director, Department of Ecology, State of Washington, Olympia, Washington.

Nitrogen Supersaturation in the Columbia and Snake Rivers. July, 1971. A summary report by the Environmental Protective Agency, Region X, Seattle, Washington.

An Economic Evaluation of the Columbia River Anadromous Fish Program. February, 1969. Findings of Jack A. Richards, Fish and Wildlife Service, Bureau of Commercial Fisheries, Regional Office, U.S. Department of Interior, Portland, Oregon.

Effects on Fish Resources of Dredging and Spoil Disposal in San Francisco and San Pablo Bays, California. November, 1970. A special report prepared by the Fish and Wildlife Service, Regional Office, U.S. Department of Interior, Portland, Oregon.

Fish and Wildlife Habitat in Relation to Reclamation of Tidelands and Marshes in San Francisco Bay Area, California. October, 1963. A statement prepared by Fish and Wildlife Service, Bureau of Commercial Fisheries, Regional Office, U.S. Department of Interior, Portland, Oregon.

Southern California Estuaries and Coastal Wetlands Endangered Environments. Undated. A statement by Bureau of Sport Fisheries, Regional Office, U.S. Department of Interior, Portland, Oregon.

California Senate Report on the Baldwin Channel. October 27, 1971. Prepared by the Natural Resources Committee of the California Senate, Sacramento, California.

Public Hearings on Baldwin Channel before the California Senate, Sacramento, April 23, 1970. A transcript.

Battle for the Sangamon (second edition). July, 1971. Prepared by the Committee on Allerton Park, Champaign, Illinois.

Analysis of the Potomac River Basin Report of the District and Division Engineers, Corps of Engineers, U.S. Army. April, 1963. A study re-

leased by Anthony W. Smith, President and General Counsel, National Parks and Conservation Association, Washington, D.C.
Report on the Connecticut River Basin. February 11, 1970. Prepared by the Connecticut River Watershed Council, Easthampton, Mass. and sent to the U.S. Army Corps of Engineers.
U.S. Congressional Documents (available from Sup't of Documents):

House Documents	Congress	Session
643	80th	2nd
170	85th	1st
369	90th	1st
Senate Documents		
213	70th	2nd
115	71st	2nd
6	85th	1st

Index

222